THE LOTTERY'S MAGICAL EFFECT

THE LOTTERY'S MAGICAL EFFECT

THE CONTEST THAT CHANGED THE WORLD

David Singer

© 2018 David Singer
All rights reserved.

ISBN: 978-1724455789
Library of Congress Control Number: 2018909127

Disclaimer
This is a work of fiction. Names, characters, businesses, places, events and incidents are either the products of the author's imagination or used in a fictitious manner. Any resemblance to actual persons, living or dead, or actual events is purely coincidental.

DEDICATION

I dedicate this book to the many victims of wars, poverty, crime and those who struggled to survive and overcome the strife and emigrated to different lands. Their main objective being to make their children's life much better in a safer world.

TABLE OF CONTENTS

World Good News Association (WGNA)
    The 18 Point Contest xi
Preface. xvii

Chapter 1   The Super Surprise of my Life 1
Chapter 2   My Feelings 4
Chapter 3   Our Children and Family 6
Chapter 4   Nelly's Thoughts 9
Chapter 5   Our Meeting 11
Chapter 6   The Preliminary Plan 14
Chapter 7   Our Recruitment Proposal Begins 20
Chapter 8   Advertisement in Major Newspapers
                   of the World. 23
Chapter 9   Philosophy of the Participants 25
Chapter 10 Luck Strikes Again 27
Chapter 11 Spokespeople. 32
Chapter 12 Mike Frederick's Thoughts 34
Chapter 13 The Enthusiasm Continues 37

Chapter 14	A Strange Phone Call is Received	39
Chapter 15	The Conference Begins	41
Chapter 16	The Second Day	47
Chapter 17	The World Reacts	52
Chapter 18	Our Family	55
Chapter 19	Applications Galore	57
Chapter 20	The Russians Attack	60
Chapter 21	Betty Klash's Thoughts	64
Chapter 22	Our First Applicants	66
Chapter 23	The Members Meet to Re-affirm Goals	71
Chapter 24	Sleep Deprivation (Point #4)	76
Chapter 25	A System that Shortens Travel Time (Point #7)	79
Chapter 26	The World's Immigration Problem (Point #17)	82
Chapter 27	Nuclear Energy for Positive Use (Point #15)	85
Chapter 28	Appetite Control (Point #8)	87
Chapter 29	Food for the World (Point #3)	89
Chapter 30	A Comprehensive System to Test Compatibility Between Couples (Point #11)	93
Chapter 31	Hair Growth (Point #6)	95
Chapter 32	Education for Al (Point #12)	96
Chapter 33	The Contest Advances	98
Chapter 34	The Nobel Prize	100
Chapter 35	My Thoughts	102
Chapter 36	Not Everyone likes What we are Doing	105

Chapter 37	The Dictators of the World Call for an Urgent Meeting	107
Chapter 38	The World's Security Organizations are Alerted	110
Chapter 39	A Secret World Security Organization Makes a Plan and Gets Ready	112
Chapter 40	Russia Informs	114
Chapter 41	The Contest Advances: Its Consequences are Felt	115
Chapter 42	The South Korean President's Proposal	118
Chapter 43	The Genius Advisory Board is Formed	120
Chapter 44	Things Become Difficult	129
Chapter 45	Interpol is Called	132
Chapter 46	Mike Disappears	140
Chapter 47	The Inspector Makes a Decision	147
Chapter 48	The Contest is Unstoppable	150
Chapter 49	Dr Cantor's Thoughts	152

WORLD GOOD NEWS ASSOCIATION (WGNA) THE 18 POINT CONTEST

Contest: We need to invent and develop and perfect applications for 18 of the selected unresolved problems. We must strive to help resolve the world problems on this list.

Prize: Twenty-five million per acceptable application, or solution upon confirmation of its applicability. One million dollar and up to five million-dollar prizes for special categories will be available in profit sharing by-lines. Each different successful development idea or invention will receive an individual prize.

Judges: the WGNA association members: to be named, mostly Nobel prize winners and successful past heads of states, successful business, and education leaders.

Contest Participants: Expert participants from every country in the world.

1. To the participant or team of participants, that develops a method of **Climatic Control,** to stop a hurricane, or lower its destructibility power. This also includes, tornados, earthquakes, earth warming effects, floods and forest fires. This point will also include the possibility of someone developing a way to control the climate by fixing it to the area's necessities. For example; The deserts, like the Sahara, have almost no rainfall. If someone develops a way of changing that imagine the benefits it will bring to all the inhabitants of that water-deprived area.
2. To the team that develops an application or system to better **control the traffic in the main thoroughfares in all the cities in the world.** That includes, airports, on land, air, and sea. Everywhere in the world. Can you even imagine how beneficial that will be for families, for industrial production, against pollution in the environment?
3. To the team that resolves the lack of food in the world, a method using intelligence, to **organize food development and production** in and for, all countries where **food** and **water** shortages are present. This includes making clean water availability everywhere in the world. This will make the world a better and more humane planet.
4. **Sleep deprivation control**: To the team that develops an application or system that will permit people who cannot sleep to resolve their problem with a drug free

system. Many won't understand this point until they grow older and see how important it is.

5. The definitive implementation of **cures for cancer**, heart disease, infectious diseases, hereditary disease, and all illnesses that are a threat to humans and humanity. This is easier said than done. However, look at the incredible advances in the last 25 years. I predict that in much less time this can be a reality. It will be expensive, and initially it will cause a lot of tumult in many industries, but the results will talk for themselves.

6. A safe system of **hair growth** for men and woman of all ages. I know this point will be criticized, because of its total lack of importance to the world. However, it might be very important to a very small portion to the world's population and that makes it a point to ponder. Little things count for some.

7. A system to **shorten travel time** both on the road, air, rail, and sea. This point is very relevant for the near future. It will bring the world's countries closer faster.

8. A safe and easy method of **appetite control**. We have too many obese citizens in this world. Upon controlling this problem, many of the illnesses obesity generates will diminish.

9. A force that will **eliminate those seizing governmental power** without totally transparent democratic principles and elections. (we can't have one ruler be a menace to the world).

10. A **United Nations body that works,** for all races and religions, equally. We cannot have ineffectual leadership permit civil wars and the creation of personality cults that stay in power forever to detriment to their citizens will.
11. A **love app**: whose development will inform future couples of their compatibility through genetic and statistical data. This way they will know better if they have chosen the right spouse. This will eliminate or diminish the terrible divorce statistics and contribute to lowering the abandonment of children and their abuse in world.
12. A system of **Universal education** for all, where everyone, all inhabitants of the world, have the same possibility of achieving the best education.
13. **A world electoral system** that functions with perfection and is administered by honest organizations. Where there is no possibility or doubts as to the final results in the elections in all countries, states, cities and this applies for every elected official in the world.
14. A **world space exploratory council**, where all space investigation, and exploratory travel, is performed together and not by individual countries. Space should be neutral, and space should not belong to any one country or power in this world
15. Definite **elimination of all nuclear** and non-nuclear weapons, (including chemical weapons), in every country in the world.

16. A **Supreme World Court** integrated by the 25 largest countries in population and empowered by Noble prize recipient judges. They would immediately rule on conflicts between countries, and any other important matter in the world. The special forces that will be constituted, will enforce their rulings, immediately after the decisions are made
17. **An Immigration regulatory world council**, where countries that need more immigrants will ask countries who have an overpopulation of available citizens for migration. All this will proceed on a voluntary basis only. This point will make possible the re-equilibrating of overpopulated and underpopulated areas in all the countries in our world.
18. A special mixed commission that will have the know-how, and power **to resolve the mid-eastern** conflict. If many of the above points begin to find solutions this point might become more viable.

PREFACE

The story I'm about to tell you deals with luck. Luck is basically success or failure that occurs by chance, rather than by one's actions. In my case, I guess it was double or triple luck that gave way to all the happenings that I'm going to relate to you.

My family lived in war torn Europe during the 1940s. My parents were born there in different cities of Romania. Both went through harsh and terrible situations, the loss of close relatives, poverty, malnutrition, and an array of psychological situations that could not and shouldn't be forgotten by me and the world's future generations. It is not the purpose of my story to relate the stories of the Holocaust and how it affected Europe. I want to make you understand the quest for reversal of fortunes that pushed my parents to their limits, in making the life of their children as prosperous as possible. I want you to understand the motivation behind our story. What made them excel in everything they did, or attempted to do, as seen by me, and my siblings.

After my parents experienced separations and re-encounters, suffering, and depravations, and just before the war ended, I was born. The main objective of all war survivors was to get out, as soon as possible from wherever they were. The situation was difficult. My parent's luck began when they caught a ship going out of the port of Constanta on the Black Sea. We ended up in Milan, Italy, right after the allies liberated that country, and everyone seemed to be a refuge. Everyone searching where to go, everyone searching for food, for the day, and for somewhere to sleep. Everyone hoping to find someone they could recognize and lead them to their lost relative somewhere in the world. When I asked my father how he survived, he told me that he and my mother took turns in looking for cigarette butts on the streets and plazas. Then they would eliminate the burned portions and retrieve the unused tabaco and then, re-roll them in small pieces of newspaper. They then sold the re-furbished cigarettes the next day. With that, they bought whatever food was available. This helped them survive.

My families' principal objective was to get to Israel. At that time the British occupiers did not permit entrance of more refugees, so they were lucky again and got on a ship that left them in Barcelona, Spain. From there, a ship transported them to Uruguay, and then, somehow, they ended up in Paraguay. In this country they lived for a while, and re-united with one of my father's brothers. They continued to live in total poverty and soon after that, Paraguay went into a civil war. They had little choice but to move on to Argentina, where Eva Peron, the President's powerful wife, welcomed all refugees and legalized them. This caused us to become Argentinian citizens.

Buenos Aires, the capital of Argentina during the end of the 1940s, was full of European refugees, many of them Italian, and of Jewish decent, from all over northern Europe. My father became a tailor's assistant and my mother helped out doing door to door sales of whatever she could find to sell. For several years that was the way of life. They lived on a day to day basis and sent me to a public school. I can't forget many things from my formative years. Some of them are -- the milkman coming around every morning at 6 am. My mother running down with a small pitcher to milk the cow herself for a few pesos. The white frock I wore as a uniform. My activity after school of butterfly catching. I remember running after them with my only toy, a butterfly net. Life was predictable. There was no fear here, there was no war. All was fine here, and the fact that we had survived is all that mattered. At the end of 1950 my mother became pregnant and a flurry of events took place.

Israel had by then become a nation and opened up a consulate. People flocked there to attempt to see if lost relatives could be located. We were lucky again. My maternal grandparents finally had found a way of locating us. They had migrated to the Dominican Republic. In those days the President, Generalissimo Rafael Trujillo Molina, had permitted Jewish immigration and my grandfather, a very competent businessman, had opened up a mirror factory in the then capital Ciudad Trujillo. He sent us ship tickets and after my brother was born, we headed for the island of Hispaniola. That is the island that the Dominican Republic shares with Haiti. My parents started to work in sales. My father in my grandfather's shop, and my mother as a door to door vendor of the article of the moment, nylon stockings and tablecloths from Taiwan. They made sufficient money to get by. My father was very careful to save as much as he could for the future. Every penny counted in those days. Life went on, and things continued without change. My mother had become addicted to saving some of her hard-earned money and every Sunday she purchased lottery tickets with some of her savings. I remember my father used to get very angry at her and would always remark "you waste money uselessly." One Sunday his anger completely disappeared: that afternoon when the lottery winning numbers were announced my mother showed him that she had won the first prize a sum of about $45,000 United States dollars. That addiction was the beginning of an incredible change in my families' life. We had suddenly become wealthy. This occurred in the early 1950's. Only very wealthy people had that kind of money. My father was able to apply and obtain a United States visa, and flew to New York, we had relatives there. He put half into a saving account, and the other half was used to purchase merchandise for his future store in the Island. That was the beginning of his successful retail and wholesale business that prospered as the years went by. My mother continued buying lottery tickets and won a second time. This time a lesser amount, but

THE LOTTERY'S MAGICAL EFFECT

she passed on her incredible luck to her future generations. I was sent to Brooklyn, New York at a very early age to study in a religious school. I wasn't very happy, but I could not complain. I was too young to disagree with my grandfather's and father's decisions.

I met my future wife while in college in New York and went home to tell my parents that I wanted to get married. I was 19 at that time, I was very independent. I lived most of my life alone, in boarding schools or in rented rooms. The day after I arrived in Santo Domingo to talk to my parents about getting married, a civil war began. My family happened to be living in the area occupied by the Cuban backed rebels, and we went through several days of constant fear of being shot by one of the war faring sides. Luck was on our side. My father was able to reach a deal with the rebel commander, and we were escorted to the government held side of the city. In those days the President of the United State was Lyndon Johnson and he sent help to all US citizens. I was airlifted in a helicopter to a US aircraft carrier, then sent to Puerto Rico in a Navy destroyer. My family followed me shortly. Luck was definitely on our side.

I graduated, I got married, moved to Caracas Venezuela and became a dentist. I worked very hard, and one day I decided that I had to do something in my life to make a difference. It was easier said than done. What was I to do, that would make a difference? I moved back to the United States after 45 years as a practicing dentist. My wife and I immediately made plans to do something that would keep both of us busy, happy, and productive. We had led a very intense life. Mostly full of good things. We decided that retiring did not mean that, it was time to wait until we die. We considered that moment as the beginning of a new opportunity to bring something different to our life. The truth is that while in Venezuela we were preoccupied with work. We concentrated on bringing up our children correctly. An important element in our life as a couple was omitted. The developing of everlasting friendships and relationships. That does not mean that we did not have friends, because we did. I guess that many of them were lost over time, maybe because we catered to foreigners that did not remain long in Venezuela. People that worked in the petroleum industry, which in the beginning of our Venezuelan experience was mostly handled by US companies and their executives. These were easily attracted by us, mainly because of my profession. Everyone wanted to be friends with a friendly English-speaking dentist in a foreign country. These people had little or no relationships. Venezuela was a temporary place for them. The thing is that all these executives and their families frequently changed jobs and moved to other foreign petroleum related jobs in other countries. In other words, these relationships did not last long. This was an ongoing and repetitive story. The new ones, which took their place became our new friends and then left also. Only very few of these relationships lasted and still exist. When the Petroleum industry was nationalized in Venezuela. All these executives ceased coming, and then we learned of the need to change and search for people with a steady relationship within the country. We were members in several of the community clubs, both in the city as well as in the beach areas. We were founding members of a Rotary Club. We were friendly with our children's friends' parents. We also had relatives, and we were relatively close to most, but not all of them. The idea came to our mind that in old age, if we were lucky to reach that stage, we had to make changes. We knew that in later years our children would be occupied with their life. We as parents would definitely not want to interfere and depend on them for our very necessary social life. We knew that only with the ability of friendship cultivation, would we lead a full social life. Which usually brings moments of joy and happiness. Much different from the joy felt about the achievements of one's children, and grandchildren, in the different stages of life. I was still in search of making a difference. I thought of many possibilities. I had been a Boy Scout when in school in Brooklyn. I had been a Rotarian for over 25 years in Caracas. I was a person who wanted to help and serve my community. I always participated in charitable events. That was not sufficient for me. Then, I decided that I wanted to write a book. I started to write. The first thing that came to my mind was a children's book of a story that I had been telling my children and grandchildren. One of my granddaughters said one day, "Grandpa, why don't you write the story you used to tell us all the time." That's what I did. I wrote it in Spanish. "El Barco Mas Grande y mas Lindo Del Mundo," or "The Biggest and more Beautiful boat in the World." Later I changed the name of the book to, "The

Contest for the Most Beautiful Princess in the World." This started me off, and it was like a bullet fired accidentally. I couldn't stop writing. My first fiction book, "The Success of Failure," turned into a second one, "Body Shapers Dream Team," then followed, "The Mind Reader," "The Lottery's Magical Effect," "The Unusual History of my World," "Cosimia" and "The Dentist."

CHAPTER 1

THE SUPER SURPRISE OF MY LIFE

It was an ordinary hot and very sunny day in Miami, when I went into the supermarket to buy some fruit and vegetables. I saw a large crowd in a line, and, I went up to the last one in the line and asked him, why are there so many people in this line? Are they giving away something for free? He said no, tonight is the big Powerball Lottery. I decided to stay in line and spent $20 dollars on several tickets, and continued shopping.

When I got up the next morning, I had completely forgotten about the Powerball tickets I had purchased. Then, I switched to the television news channel and heard that the lucky winning ticket of the lottery had been sold in the Miami area, and that one very lucky person had won a one billion and three hundred-million-dollar Jackpot. $1,300.000.000. This was the largest jackpot accumulation ever won anywhere in the world. I mentioned the incredible sum to Nelly, my wife, and she said, "You should try your luck. Your mother won twice." I told her about the $20 I had spent yesterday on the tickets. She told me to give the tickets to her. No, I'll do it, don't worry I'm busy writing. She kept on insisting. I started googling for the winning number and then, I almost had a heart attack! I compared the winning number to one of my tickets 15 times before calling my wife. She thought I was joking with her, when I said, "We won, come here look at this ticket." She came over, triple checked and also almost had a heart attack.

We decided to sit down, calm down. I took a picture of the lottery ticket with my iPhone. Then, I photocopied the ticket, just in case. We decided that we had to plan on how we were to manage this. We can tell our children later. It won't make a difference. We need to keep this quiet. If we make our winning public, we will be the center of attention of every newspaper, TV station, magazine in the world. They just said it was the biggest Jackpot in history. Every thief and everyone that has something to sell will be after us, and also our kids, and family. We have to keep this a secret from the press. We can only share this with our banker, and the tax people of course. I don't think we need a lawyer, not yet, at least.

We really did not have a relationship with a banker. Nelly and I went into one of our local banks and waited to talk to one of the bank officers. After one hour of waiting we decided to go to a different bank, we were too nervous to continue to just sit there. This time we went to one of these banks that are open every day of the week. We had a small account, that we didn't move frequently. We were lucky, no one else was waiting. I asked the officer for the bank manager and he was very nice and took us to his office. We told him that we had a very important sum of money that we wanted to deposit and wanted to keep this a total secret. We saw that he started to look at us as if we were strange and crazy. We then decided to tell him that we were the winners of last night's jackpot. He couldn't believe it. I showed him my photocopy of the ticket. He opened his computer to search for the number, and he confirmed it. He told me he would have to check with his head manager at the head office to see how to proceed. We then waited. After multiple phone calls, we finally handed over the ticket to the manager. We had to sign all kind of forms, but we were not guaranteed secrecy. This was too big a deal to maintain in secret.

CHAPTER 2

MY FEELINGS

The first thing that came to my mind was that I would use the money to create something that would make a difference. My dream now might come true. Money talks, and now I have money. I will make myself heard. There comes a time in the life of an accomplished person when he must make a decision. Should that person leave things as they are? Should that individual perform one last attempt to make a difference in this world, or in his community, or in his environment? Or, should one be content with what he or she has done or achieved or accomplished in life until now? The natural question that has to be made is, "What have I done up until now? Is this enough? What have I really achieved? Do the people I care about, know of what I have had to do to get to this point? Do they consider this normal or expected? Am I satisfied? Do I think I can do something much better?" If the answers to these questions that I ask myself are not clear, then it is time to begin. If age health and time permit me to start thinking of what I can do to make a difference. Why do I have to make a difference? Well I need to be creative and I want to share this feeling. I want my children, my grandchildren, and eventually my great-grandchildren to feel the need to imitate me. I want to create something of great utility and show them the never-ending feeling that I want them to have. I really think that if they imitate the feeling I have, they will be successful in anything they attempt in life. Well, does this mean that I have been successful in life? My answer is that I have definitely not been unsuccessful and that means that I am Ok. I am doing well, but I definitely think I can do more. Or at least one more thing that will make an even bigger difference in our and their lives. Every time I do one more thing, or I finish doing something, I immediately feel the need to look and search for something more creative. I now had money, and that means that I have power. I have power to make myself heard and understood -- let's be honest, make myself respected. People tend to respect and look up to someone with power. Someone with money and with strength. This gave me a new back up, I now had all these.

CHAPTER 3

OUR CHILDREN AND FAMILY

A fter the paperwork was taken care of, the bank manager assigned to assist me advised that it would be in our interest to hire a tax accountant and a financial advisor to handle our new bundle of problems. We needed information on how to dispose of the fortune with care and intelligence.

Nelly decided to call our children and told them that we needed to meet with them urgently and that we needed them to bring their children, our grandchildren, to that meeting. It was a matter of life and a much better life. We added, that the meeting would be in New York City in The Peninsula Hotel. We included an electronic first-class airline ticket for each one, and we were sure they understood that all the questions that they might have about this sudden and unusual and very expensive paid request would be answered on that weekend at the Peninsula. We promised to each other, that under no circumstances would we divulge the reason behind our request. We knew that many things might go through their mind. They would consult each other about the motives behind this. They knew we would never spend money unnecessarily, and never go to such an expensive deluxe hotel. We never used first class airline seats, and they knew it. They might think we were into a demented state and were spending our very hard-earned money because of something as fatal as being fatally ill and wanting to get the family together one last time. We were immediately called by our kids. Some of our grandchildren said they couldn't come because of an exam, or other plans, or whatever. Our answer to them was that if, God forbid, they needed to assist at one of our funerals, would they also find an excuse and not cancel their date or bring their material to study here to New York City? That statement changed their attitude and they felt possible grave circumstances surrounding our urgent request. The same applied to our children's wives and husbands. They had made plans, and some wanted to be excused. My answer was the same I gave to my grandchildren. That made everyone uneasy and nervous. Some called our close friends to ask about our strange behavior. They were of course unsuccessful, because we maintained a continued and very close relationships with everyone we knew, and they reported no change in our ways.

We also decided to meet with our close family, brothers and sisters, back in Florida a few days later. We made a list of relatives that we knew were in a needy situation, and some friends back in Venezuela, who we knew well, and we knew could use a hand. We also decided to include a charity fund earmarked for some of our alma maters, and places of worship, and institutions that served the less privileged.

CHAPTER 4

NELLY'S THOUGHTS

I have had a great life. I will never forget the day I met my husband. I really didn't formally meet him that day. I knew that he wanted to meet me, when he kept on coming by while I was taking on sun in that beautiful spring day in the lawn of our university's football field. He attempted to make himself visible but didn't know how to approach me. These moments continued for several weeks until one day he had the courage to come up and ask me if I was from Argentina. Someone had told him I was a foreign student from a South American country. After I told him I was born in Romania but grew up in Venezuela, he found a similarity in our background. He told me he was also born in Romania. After a few more moments we found out that we knew each other's relatives. Two years later we were married in New York. We had three children and went through many difficult economic situations until he started working in his own dental practice. We married very young and we think that the fact that we were more mature than most people our age was an important reason for our successful life. Life with a dentist is not easy. Dentistry is a profession that creates stress. People who go to the dentist are usually very stressed. A better way of expressing myself is that few people who go to a dentist are not scared of the procedure they will be subjected to at one time or another. This very stressful environment causes stress on the attending dentist. The dentist should leave the stress in his practice and not bring it home. That unfortunately is not always the reality. We are humans and can't switch off our problems in one environment and become problem free when we arrive in another. He did a good job of keeping business and professional problems out of the house, but he was human and concerned and it wasn't always easy. The years went by, we enjoyed our children's achievements. Time flew. They finished their university studies, married, had their children and then dispersed. Our first ended up living in Bogota, Colombia and then in Santa Barbara, California. The second in Palo Alto California and our youngest in Miami Florida. When the political situation in Venezuela became very difficult, and safety became a serious issue, I convinced my husband to sell his practice and retire. He had worked more than 45 years and we needed some time for ourselves. He always said he wanted to travel and visit the world, something we never had time for. He listened to me and we moved back to the United States. Many times, he talked about doing something that would make a difference. I agreed with him and promised to help and participate in everything. I still loved him very much. *I knew he felt the same*

CHAPTER 5

OUR MEETING

Our children and grandchildren arrived on that magnificent, sunny, Friday afternoon. We made sure they were all on the same floor, and we had the hotel deliver the exact things that each one of them loved. Their favorite chocolate, their favorite drinks and magazines. We kept away. We wanted to create a type of camaraderie between them that had existed at one time, but since all of them lived in different places and rarely met, it disappeared little by little, as all things that are far away. We informed everyone that we would meet them for the traditional family Friday meal for Shabbat. We had arranged everything with excellent caterers, perfect for this type of an occasion.

By now they were extremely curious, nervous, and full of doubts. At exactly 8 pm. we showed up dressed informally as usual. Since we had not talked about what to wear, we found that all our children had dressed as formal as you can get. Probably because we were meeting in such an elegant and formal place. They were happy to see each other but there was fear in all of them because of the totally strange attitude of their parents, who they thought they knew perfectly well until a week ago when they received our request.

"Here we are children," I said with a big smile. "Your mother and I are so glad to see you and find that all of you could come." We hugged and kissed each other with real affection. We then sat down, made the customary blessings and the meal was served.

Right after the main course meal, I got up and told them how much we loved them, and how we wanted them to be successful in life. "We want you to know that we won the lottery last week and we are now extremely wealthy. That's why we invited you to this hotel in New York. That's why we sent all of you first class tickets. You all know we have always thought you to be thrifty and save for the future, and today we have splurged. Exceptions must be made. We hope that we don't change. We hope you don't change either. We have opened an account for each one of you here. This account will be administered by a very able financial expert. He told us you will each produce a very nice income the rest of your lives if you do not touch the investments that he has chosen for you. The financial institution will be in touch with you and inform you of everything in the near future. We will also make sure that our siblings are comfortable and help relatives that we know need assistance. Your mother and I will make plans on how to use a portion of our winnings to make the life of people here and in the rest of the world easier. We don't know exactly how yet, but we will keep you informed of everything." They all shouted with excitement and let us know of how the mysterious way of getting us together here changed their lives. They started to communicate with each other a few times a day trying to guess at our intentions and we all became closer and promised to keep in touch several times a week. My wife said that she would love reunions several times a year. "Kids don't worry, I will pay for all the expenses." Everyone smiled and hugged each other at all the great news. They all started to ask questions and we tried to answer all with humor and wit. We then met with each child and grandchild individually and gave them details of what we expected of them. To maintain equality, we told them that we would control their new funds until they got used to the idea of being wealthy. We wanted them to keep on living like they had been used to. Some didn't understand why we did this, others didn't care. We did it because we care, was our answer. The next two days were full of joy and emotion and we kept busy with visits to theaters and museums. We enjoyed every second.

CHAPTER 6

THE PRELIMINARY PLAN

We met with our siblings and told them of our incredible luck, and the arrangements we had made for them and for our nieces and nephews. We wanted everyone to be comfortable in economic terms at least. We gave instructions that on a given date several institutions were to receive anonymous donations. Our tax lawyers had given us very specific instructions as to how to best handle these donations. All we know is that we created total happiness in all those who we decided to help. Naturally! All these exciting moments reminded me of the stories my mother had told me about all that she and my father went through during the war in Europe. My mind went through all those details of the German detention camp where she gave birth to me and of the incredible rescue my father planned and made possible: he saved us and many others. It made me reminiscence of the great difficulties they had in attempting to leave war torn Europe and how they went against all odds and migrated from one country to the other until they settled in the Dominican Republic. Although they are not alive anymore I had them in my mind all the time. They were fundamental in my success in life.

My wife and I sat down after having accomplished all these incredible feats. We had taken care of all our nearest and made sure they would be well off economically for a long time. Our next step was to be the real and very difficult one. We decided that we were going to search out and put together a group of specialists in many fields. Each one of them was to advise us in the best way to invest for the future by resolving ongoing problems in the United States and the world. We made a list of problems to resolve. We decided to form a company GWNA (Good World News Association It was going to be a non-profit organization, of course.). This was to be the framework for our contribution to make the world a better place. We thought that one way of finding these specialists was to make a world contest and the following are the preliminary rules.

WORLD GOOD NEWS ASSOCIATION (WGNA)

Contest: We need to invent and develop and perfect applications for 18 of the selected unresolved problems. We must strive to help resolve the world problems on this list.

Prize: Twenty-five million per acceptable application, or solution upon confirmation of its applicability. One million dollar and up to five million-dollar prizes for special categories will be available in profit sharing by-lines. Each different successful development idea or invention will receive an individual prize.

Judges: the WGNA association members: to be named, mostly Nobel prize winners and successful past heads of states, successful business, and education leaders.

Contest Participants: Expert participants from every country in the world.

1. To the participant or team of participants, that develops a method of **Climatic Control,** to stop a hurricane, or lower its destructibility power. This also includes, tornados, earthquakes, earth warming effects, floods and forest fires. This point will also include the possibility of someone developing a way to control the climate by fixing it to the area's necessities. For example; The deserts, like the Sahara, have almost no rainfall. If someone develops a way of changing that imagine the benefits it will bring to all the inhabitants of that water-deprived area.
2. To the team that develops an application or system to better **control the traffic in the main thoroughfares in all the cities in the world.** That includes, airports, on land, air, and sea. Everywhere in the world. Can you even imagine how beneficial that will be for families, for industrial production, against pollution in the environment?

3. To the team that resolves the lack of food in the world, a method using intelligence, to **organize food development and production** in and for, all countries where **food** and **water** shortages are present. This includes making clean water availability everywhere in the world. This will make the world a better and more humane planet.
4. **Sleep deprivation control**: To the team that develops an application or system that will permit people who cannot sleep to resolve their problem with a drug free system. Many won't understand this point until they grow older and see how important it is.
5. The definitive implementation of **cures for cancer**, heart disease, infectious diseases, hereditary disease, and all illnesses that are a threat to humans and humanity. This is easier said than done. However, look at the incredible advances in the last 25 years. I predict that in much less time this can be a reality. It will be expensive, and initially it will cause a lot of tumult in many industries, but the results will talk for themselves.
6. A safe system of **hair growth** for men and woman of all ages. I know this point will be criticized, because of its total lack of importance to the world. However, it might be very important to a very small portion to the world's population and that makes it a point to ponder. Little things count for some.
7. A system to **shorten travel time** both on the road, air,

rail, and sea. This point is very relevant for the near future. It will bring the world's countries closer faster.

8. A safe and easy method of **appetite control**. We have too many obese citizens in this world. Upon controlling this problem, many of the illnesses obesity generates will diminish.

9. A force that will **eliminate those seizing governmental power** without totally transparent democratic principles and elections. (we can't have one ruler be a menace to the world).

10. A **United Nations body that works,** for all races and religions, equally. We cannot have ineffectual leadership permit civil wars and the creation of personality cults that stay in power forever to detriment to their citizens will.

11. A **love app**: whose development will inform future couples of their compatibility through genetic and statistical data. This way they will know better if they have chosen the right spouse. This will eliminate or diminish the terrible divorce statistics and contribute to lowering the abandonment of children and their abuse in world.

12. A system of **Universal education** for all, where everyone, all inhabitants of the world, have the same possibility of achieving the best education.

13. **A world electoral system** that functions with perfection and is administered by honest organizations. Where there is no possibility or doubts as to the final results

in the elections in all countries, states, cities and this applies for every elected official in the world.

14. A **world space exploratory council**, where all space investigation, and exploratory travel, is performed together and not by individual countries. Space should be neutral, and space should not belong to any one country or power in this world

15. Definite **elimination of all nuclear** and non-nuclear weapons, (including chemical weapons), in every country in the world.

16. A **Supreme World Court** integrated by the 25 largest countries in population and empowered by Noble prize recipient judges. They would immediately rule on conflicts between countries, and any other important matter in the world. The special forces that will be constituted, will enforce their rulings, immediately after the decisions are made

17. **An Immigration regulatory world council**, where countries that need more immigrants will ask countries who have an overpopulation of available citizens for migration. All this will proceed on a voluntary basis only. This point will make possible the re-equilibrating of overpopulated and underpopulated areas in all the countries in our world.

18. A special mixed commission that will have the know-how, and power **to resolve the mid-eastern** conflict. If many of the above points begin to find solutions this point might become more viable.

CHAPTER 7

OUR RECRUITMENT PROPOSAL BEGINS

Today we have the Internet, through which we can search and find whatever we need, so before starting, we needed to make a list of specialists that can help us select the contestants and the judges.

The natural and most reliable sources will be the top research institutions in the United States. We can access the top minds in the United States and the rest of the world through a contest. My wife and I think that once we make this public all the top learning institutions will want to participate and compete for the incredible prizes and the real contributions these results would produce, including a better and safer life for the inhabitants of this world. To add weight, status, and credibility to this project we must appoint a very well-known apolitical, and distinguished personality as the head spokesmen.

THE LOTTERY'S MAGICAL EFFECT

We both decided to see if we could contact some of the wealthiest men and women in the world and invite them to participate in this incredible global undertaking. They will help us with the selection and managing of this super world project. Let's contact Larry Page of Google, Mark Zuckerberg of Facebook, Jeff Bezos of Amazon. Warren Buffett of Berkshire Hathaway, Oprah Winfrey, media magnate Carlos Slim, of America Movil. Jaime Dimon, from Chase, Elon Musk, from Tesla. Jeffrey Immelt, from General Electric, Lloyd Blankfein, from Goldman Sachs. Doug McMillon, from Walmart. Jack Ma, from Ali Baba, Akio Toyota, from Toyota. Sergio Brin, from Google, Charles Koch, from Koch Industries. Tim Cook, from Apple, Rupert Murdoch from News Group, Jim Yong Ki, from The World Bank. Michael Dell, from Dell, Michael Bloomberg, from Bloomberg. Carl Icahn, from the Icahn capital management group and Bill Gates, the Microsoft founder and the Bill and Melinda Gates Foundation. To name a few of the personalities that will want to join us, we think, in making this project a reality.

As the head spokesman I will attempt to contact Michael Kirk Douglas, a world renown actor and producer. He would be ideal in every respect. My wife loved that idea.

For our idea and our company to become a reality we must invest in advertising our commitment, our purpose, and our request for the above-named dignitaries to get motivated and involved, and ultimately, contact us. This is also applicable to anyone in the world that wants to contribute to make this contest a success. We are also open to all ideas and modifications to our original proposal and welcome additions and modifications that will make this world contest even more competitive and successful.

CHAPTER 8

ADVERTISEMENT IN MAJOR NEWSPAPERS OF THE WORLD

We contacted a major international advertising firm and asked them to immediately start the campaign. First, we took out full-page advertisements in *The New York Times*, *The Wall Street Journal*, *The Chicago Tribune*, *The Washington Post*, *The Los Angeles Times*, *The Boston Globe* and *USA Today*, in the United States. *Yomiuri Shimbun* in Japan, *The Times* of India, *People's Daily* from China, and *The Daily Mail*, *The Sun*, and *The Guardian* from United Kingdom, *The Chosun Ilbo*, from South Korea, *The Australian*, from Australia, *El Pais*, from Spain, *Le Monde* and *Le Figaro*, from France, *La Corriere Della Sera*, from Italy, *Die Welt*, from Germany, *The Toronto Star*, from Canada, *El Clarin*, from Argentina, *The Jerusalem Post* from Israel, *O Globo*, from Brazil, and *Pravda*, from Russia. Plus 100 other major newspapers. We knew that all the rest of the news media and Radio Television and Internet high tech media would immediately comment on this advertisement blitz. The advertising firm did not think anything more would be necessary. The interview requests by all media would follow and we needed the spokesman capable of explaining the very clear purpose and contest rules and prizes. We rented an office in New York, where we had moved to, and had it staffed by an employment organization. We had everything ready to begin our goal of "making a difference." The Lottery's magical effect was starting to be felt.

CHAPTER 9

PHILOSOPHY OF THE PARTICIPANTS

When one becomes rich or is born with wealth, many things in life become easier, but mostly, let's call them multimillionaires, know that they are different. They tend to be different in their approach to many things, especially the material ones. The ones money can buy. Once these men and women reach a plateau, in their life, where they have gotten used to living their lifestyle and have all the necessities, and all they had wished to buy, own and use. There comes a moment that all want to do something for others, or so it should be. They should all want to help their society, their schools, and churches, their people, and all people who think like they do, or those who can use their help.

The participation in this contest will not only achieve all their possible wants, and wishes, but will give them a partial control of so many things in the world that need solutions. With the participation in our group, they get to be part of possible solutions for the major problems in our world. The participation in our group "makes them make a difference in today's world problems" because our contest will achieve major solutions, we have no doubt.

CHAPTER 10

LUCK STRIKES AGAIN

It was early in the morning when our new financial advisor, Mike Frederick, sent me a message to call him back. That was strange, since we had agreed to communicate only in case we had to make a very important financial decision. I called him, and he explained that he had something to discuss with me that might be very helpful to my plan. He was willing to come over to our house or meet anywhere. He needed to tell us privately of his discovery. My curiosity couldn't wait, so I told him to come right over.

The excitement in his face and voice were evident. This man was the chief financial officer of one of the principal banks in New York. He was once on the list of candidates to preside the board of directors of the World Bank and also, the International Monetary Fund (IMF). Mike had decided that he personally wanted to take over the financial aspect of our plan. He sat down and got down to business immediately. He was not the type to get into unnecessary small talk.

"Dr. Cantor, you won't believe what incredible news I have today. I had an appointment with one of the most important clients that our bank has. He has reached the point in his life where he also wants to make a difference. He asked me for my advice. Many of the bank's clients do that. He wants to get involved in something that will make an incredible difference for as many people as possible in the US and the world. Immediately your contest plan came to my mind. I have heard of many possible ideas on how to do something for society, but never have I heard of such an incredible idea as a world contest for solutions to some very important unresolved problems of today. I then told him that I might have a plan to share with him. I did not tell him anything without your permission. He is a well-known billionaire, and he wants to contribute whatever is necessary of his fortune, this includes his organizations, and his staff. He also has an intimate relationship with the top people in the world of industry. He might be able to contact other very wealthy contributors." I told Mike that this is no secret, you can share our plan with everyone. As soon as someone joins our group it's their plan also. I will retain principal ownership of the contest, let's call it the Association just to maintain our original idea and avoid dramatic changes. Remember we are a non-profit. We will accept all ideas and alteration of the 18-points, if necessary, to add other points to the contest, whenever the majority of members consider it appropriate. Mike was extremely happy with my explanation and acceptance of his proposed member. "So, Dr. Cantor, from what you just said, I am authorized to expand my personal search for "possible useful members of the Association? Dr. Cantor, when can I schedule a meeting with my man." "As soon as you want." "Mike, have you started to interview executives for the General Manager spot in our new office?" "Yes, I am waiting for her response. I offered the job to the former president of the World Bank, Ms. Betty Klash. I do not have information as to the salary I can offer. Mike, I trust you, and we know that you will only get the best. So, ask them how much they want to earn. Money should never be an obstacle to anything we do."

The next day we met Harry Tower, former Chairman of the Board of one of the largest Pharmaceutical firms, and, I might add, one of its principal stockholders. Mike had explained everything of our vision, to the last detail. He wanted to collaborate with one billion dollars to start. That alone doubled our pot. He also wanted us to give him a job. He did not want to be a silent partner. I told him that if he wanted to, he could have the responsibility of contacting all the important men and women we hoped would contribute to our plan. "Get them together as soon as possible. My wife and I will dedicate 100% of our time to get this project off the ground." "Dr Cantor, you have no idea how happy I am. I knew that I could count on Mike. Your plan is God sent." "Dear Mr Tower, you are now an equal partner in our Association. Everyone who joins us will be."

That same week Mike informed us that Mrs. Klash had accepted the General Manager position. She said, she did not expect to be paid. "It will be a great honor to serve in this capacity." She wanted to hire her own assistants. Mike had given her a green light on everything. Mr. Tower had arranged a meeting with three incredible contributors. One pf the principal stockholders in Sweden's phone enterprise. The past CEO of Germany's principal automaker, and Mr. Kitza, a Japanese conglomerate head. Mike had contacted one of the heirs to the well- known US cosmetic industry and five others who wanted to speak with me personally. After our conversations and their donations, the money my wife and I won in the lottery seemed to be seed money for all these wealthy people, so they could to follow my lead.

I had attempted to contact the man that I wanted as spokesman. We had not been able to contact his representative. My wife's friend suggested we contact one of the past presidents of the US, she was sure all would accept. I told her that I did not want to involve politicians. This is an affair for the world. We do not want politics to create different alternatives that might be biased. She understood. The idea of Douglas being our spokesman really appealed to me. He was an incredible actor, winner of a number of accolades both competitive and honorary. Everyone on the world scene would recognize him and respect him.

The next morning, I received a call from the Wall Street Journal, asking me for an exclusive interview. I accepted of course but said it could not be exclusive. The New York Time and Le Monde followed and so did many others. I needed a press secretary as soon as possible. Mike had spread the word and the networks were after an interview also. I decided to move from my home office to the associations office in New York. There we had the new team initiating their organizational duties. Mike Fredericks advised me to be careful. He implied that my wife Nelly and I were not safe. Our names and pictures were all over the newspapers and newscasts and, "There are people with bad intentions everywhere."

CHAPTER 11

SPOKESPEOPLE

A fter the first week with our contest idea in full swing, I saw that we needed several spokespeople, not only in New York but in all the principal cities in the US, and some European and Asiatic countries as well. I contacted our new members and asked each of them for a suggestion. The answers were immediate, and all wanted to hire the top of the line, well-known and respected candidates in their cities. A meeting of all these new volunteers was set up in The Jacob Javits Convention Hall. Mike and I were to be the principal speakers. We originally planned on 200 attendees, then our General Manager called us to say that she had seen the need to add 800 additional seats and had more requests but decided to stop. She said that two ex US Presidents wanted to speak also, but she did not have slots to give them. All major networks were to be present. The US Vice President called to congratulate me on this incredible initiative, in the President's name as well, of course. Betty Klash called us again to ask if she could make it a two-day conference, because several of our members solicited speaker times and various countries contacted her, asking for speaker time for their presidents or prime ministers. We decided to accept her suggestion. I thought it might be a good idea to let the world press receive the planned rules and regulation and our 18 objectives at the end of the conference. Mike commented that China was the only country that had not shown interest. I corrected him. I'm sorry, I forgot to inform that this morning I received two calls from China. One from a government official asking for space for the minister of science and technology, and the other call from the principal stockholder of Ali Baba himself. He said he wanted to contribute with money and personal time, he told me he had several suggestions to make. I welcomed his interest and told him to contact Betty. We were all so busy that Betty and I decided to give up on trying to get Kirk Douglas as our main spokesman. He was simply unreachable.

CHAPTER 12

MIKE FREDERICK'S THOUGHTS

I remember the first time I met Dr. Cantor and his wife Nelly. They were a very impressive couple had been married for over 50 years and still held hands in public, and did not look their age, especially Nelly. She was a thin good-mannered, and I might add, sexy looking lady. Dr Cantor looked like a respectful elder. Both were full of enthusiasm and ready to tell me their story. They were referred to me by the manager of one of our Miami branches. Managers never refer clients to me directly. They always go straight to one of the many wealth managers in my department. This particular manager, I knew personally. He had achieved excellent results in our internal marketing objectives. He had given a talk to our managerial conference last year and was congratulated by all the attendees for the originality of his thoughts and actions after a hurricane had devastated his area. The Cantors explained their luck, their plans and their objectives. They really impressed me. I am not one to be easily impressed by anyone. I have been in the investment banking business for many years and have been very successful. I had a lot of responsibilities in my bank and with many other organizations. I was on the board of several corporations and was continuously asked to join others. To be honest, I have been thinking of retiring. Then the Cantors come in with the greatest idea I have ever heard. All I knew about them was that Dr. Cantor was a retired dentist who had practiced in Venezuela and that they had recently won the biggest prize ever paid out by the lottery in the US. He called his idea "The Lottery's Magical Effect." I was impressed that they had chosen our bank, as the bank to handle their new fortune. I became totally immersed in their project. It was like a magnet had penetrated my mind. The world contest was such an incredible idea, and their objective was clear, "They wanted to make a difference." And I knew, if they permitted me to join them, that I would make that difference that I wanted to make in the world. I got so involved that I let my assistants take over most of my day-to-day obligations. I wanted to dedicate as much time as possible to the Cantors. Since they asked me to be in charge of hiring the General Manager of the newly formed Association. I selected the woman who I hope to make my wife, Betty Klash. I have been in love with her for the last 40 years. I never had the ability to ask her to marry me and some rich man from Luxemburg beat me to it and married her. They have been in divorce proceedings for more than 10 years, and as soon as she gets the divorce, I'm marrying her. I know she feels the same about me. We have both been in silent love with each other since I can remember. I never married because of her. She was never able to have children. Her life was dedicated to problems of the world. She had served as President of the World Bank for many years. She had not accepted a new job since then. She was writing a book on the world economic outlook for the coming years and was totally dedicated to finishing it. The day that she called me to tell me she was she had finished, and I offered her the job as the General Manager of this great new plan. She accepted immediately. Initially, she accepted because that would bring her back to New York where I was going to be living now. Then after she heard the details and the complexity and the possible rewards that the contest was going to generate for world peace, stability, and for advancement in health, transportation and all the other 18 points, she became immersed in this project like I did. Like everyone who I have been in contact with. All the clients' friends and acquaintances that I have contacted have instantly become enamored with the project and thanked me for having thought of them. I had even asked the chairman of our bank's board, a very wealthy man, to join us and he said, "If Mike Frederick asks me something, I never say no." This gave me distinct pleasure to hear. He is one of the 10 richest persons in the world.

CHAPTER 13

THE ENTHUSIASM CONTINUES

In preparation for the event in the Jacob Javits convention hall, we decided to have the first meeting with the initial members of the project. We met in Mike's office and there were 12 of us. Mike and Betty, my wife Nelly, Harry Tower, Mr. Kitza, Mr. Peers of a steel conglomerate, Mrs. Puig of the Walton Corp, and four other very important and wealthy members. Mr. Towers, the first member to join our group proposed that two points in the 18-point contest should be removed. He said they were not so important for the world problems. He was referring to Point #6 and Point #8, the one about hair growth and the one about appetite control. Mr. Kitza thought that there were too many points in our plan. He added that some of them could be bundled into one point. He gave us an example that everything having to do with health should be united into one point of the plan. Betty Klash, our general manager did not agree. She said that we should keep the 18-point contest. If need be, we can substitute some of the points, but let's not make it less than 18 points, or longer than 18 points. She spoke of a psychological component of the number 18. Ms. Puig spoke of the Kabbalistic significance of the number 18. Mr. Peers spoke about the incredible world interest that the contest has generated. He said, "We have not spent any money on advertising on our project and every newspaper, every radio and television network in the world are commenting on it. Mike spoke then and let us know, that so far, "Our current members have pledged over 18 billion dollars." The money has been placed in a fund that will generate interest. We will use this fund to make contest results realities." Mike added that the 18 billion dollars is the initial pledged sum and we have an incredible list of people waiting to see the contest rules and the 18-point program. I calculate that right after the conference we will have over 100 billion or more in pledges. Mike added, that the new Chinese member, pledged three billion dollars this morning.

CHAPTER 14

A STRANGE PHONE CALL IS RECEIVED

The situation started with a phone call to the Associations office. It was an overseas call asking to talk with Betty Klash. The call was from Bangladesh, or that's what was written on the phone's screen. Betty decided to take the call expecting another possible member or an interview proposal from a newspaper. The person started to speak in British English. "I warn you that if you continue this contest charade you will all be annihilated. He said he represents a defiant organization and that his group is present in almost all the countries including in the US." He then hung up. The call lasted less than a minute. Betty had a system installed to record all conversations for future review. She called Mike, and Mike called me. We decided to consult with Harry Tower who knew everyone in Washington. He decided that we should let the local police know. They would decide whether to call the

FBI, or whatever government security group deals with calls of this type. Deep down in my thoughts I had expected a reaction from the terrorist groups of the world, but not so soon, and not so direct. At least we have been warned, we needed to be careful. Mike decided to call his bank's fraud investigation people, they had an excellent detective team. They listened to the recording and passed it on to the police and other specialized groups. We decided to double up on security for our conference, since very many high-ranking, local and foreign members were going to be present. We knew that NYC police would collaborate 100%. Special cameras were set up and intense security was to be present at the site.

CHAPTER 15

THE CONFERENCE BEGINS

B etty had extended the conference to two days. It would convene six hours every day. There were going to be over 2000 attendants. She hired temporary help to assist with the very big task of sending out personalized, non-duplicable security invitations. The seating arrangement was also very important, and we had to be very careful, not to overlook the grade of importance that each member thought they deserved. Everyone was important to us and this was a marvelous opportunity for them to bond and become an organized group. There was a main front table where I would sit with Mike and Harry Tower who we decided to name as our main spokesman. He had an outgoing personality and a very serious tone of presentation. A pre-registration application had been sent out through Internet and everyone had pre-registered before arriving. A code password was sent to each participant, so security was at its highest. We had over 500 security personnel scanning for possible intruders and suspects. Even in the bathrooms we had security cameras placed, something very unusual and probably a first-time event. The conference also had to have quality fast food restaurants especially prepared for this very special occasion.

I was to be the first speaker and the conference began on time, at 9:30 am. I told the audience of our project, our goals and Nelly's and mine dedication to the idea that we wanted to be able to show our children and grandchildren that, "We can and will make a difference." We cannot stay quiet and let so many circumstances stay un-resolved. We must search for the most able, the most daring, the most intelligent, the bravest men and women to join the contest that we will present tomorrow. These people will set the rhythm of the world's solutions by seeking to actively resolve conflicts, the medical treatments of the so far incurable diseases, the problems of the poor nations are not only their problem, they are a world problem. We all have-to help make them disappear. The differences in religions have to disappear. Everyone is entitled to their beliefs. No violence will be permitted, we have to reach that goal. Dictatorships need to be prohibited universally. All the world's organizations have-to become functional. If not, we need to change them. Laws have-to be respected by all, the poor, the rich, the super-rich, and the powerful. No one has a right to do whatever they want. Everyone has to obey the laws of the world and we should try to universalize these laws. Everyone stood up and applauded, approving my initiative and my goals. This was an outstandingly clear approval of our idea. The super wealthy present gave a clear indication that they agreed with the basic principles of our plan. They were accepting of the rule of law and equality for all without any question. Their contributions would pave the way for equality.

Mike Fredericks went up to the podium and began his talk. "I met Dr Cantor only 15 days ago, for those of you that know me, few things in life tend to impress me. Dr Cantor's plan impressed me to such a degree that I stepped down temporarily from my very prestigious and demanding job to dedicate myself totally to this unbelievable world event cause. Everyone stood up and applauded. Most of the audience knew who Mike was. The ones that didn't probably had read or seen him on their TV newscasts, when he received awards for the different benefit causes in which he was involved. His successful history in the world of finance had aided in making many of his clients past and present into multimillionaires. They all knew what he was capable of doing. They knew that the contest proposed by us was going to be successful. Mike Fredericks would not have decided to make it his project also. Mike went on to proclaim this event as the beginning of the solution to many of the world's pressing problems. You will notice that our proposal has 18 points to ponder. Some of them are super important to all the inhabitants of the world, and some are only important to a few, to only several million. However, we included them also because they are a problem to people who we know well and who we live with. They are not so terribly important, but they create a life change that will restore or make life easier and healthier for its victims.

If they resolve their individual problems, their self-esteem will skyrocket. They will be more productive, lead better and more amicable lives. Together we need to explore and look for other individualized problems that each country has. We will then, in the future include contests for the solution of their problems also. I want you to know that we will soon have over 100 billion dollars in pledges from our members. We only have 25 members right now. We hope to have several thousands in the coming weeks. The applause could be heard on the New Jersey side of the Hudson River.

Harry Tower's turn was next, he started by saying. Two weeks ago, I was depressed. I have actively participated in the United States economy and have made an incredible fortune, as many of you I am sure know. I have donated money day and night to hundreds of institutions every year since 1970. For all the money I give away. I continue to see injustice everywhere. I *meet* people that have a disease that picks them as their objective and there is no known solution. We have to find a solution for them. Of the unjust laws that abound here in our country, and we are not alone, the rest of the world has the same problems. The same unjust laws are present in Africa in Asia and in Latin America. We must make them disappear. I hear of the poverty, the violence that religious beliefs generate, of the unjust rulers that are present in too many countries. All these things, and many more, have made an impression on me. I have always heard the saying 'money talks.' I have all the money in the world and it doesn't talk for me. Then Mike, here introduced me to this now famous dentist Dr. Cantor. He became my magician, he completely resolved my depression. I studied his 18-point list and I said, this man's contest idea is the first real stepping stone to a more organized world. It will achieve more equality. Everyone who is poor, or middle class or even wealthy, has a right to the same education in every country. Everyone has a right to appropriate health care, decent housing, decent laws, decent everything. Dr Cantor's ideas will start to resolve some of the major problems of the world. Please join us, the people of the United States of America, in this incredible adventure." Everyone rose and applauded and he decreed a lunch break for everyone.

The afternoon session was very interesting. A top medical investigator spoke of the possibilities of finding a cure for many types of cancer, HIV, and new data on genetic ailments and how to approach them. He said a lot of money and excellent investigators were needed. I got up and told the auditorium that we had the money, we needed the contest to help organize the solution solvers.

The next speaker was one of our members. This man was a Texan and had the largest cattle ranch in Texas. He also owned huge acreage of land in Montana and Wyoming. He proposed that he had enough land to give away to all the undocumented citizens in the US. Let them move to one of my properties. Let them establish towns and cities there. Once they do this in a legal and democratic way, let's give them a green card and let all the law-abiding men and woman become legal. We are a country of immigrants. I will also donate whatever quantity of money they need in their process of becoming legalized. Many, but not all cheered his proposal. I thanked him and alerted the audience. Look at this man he is a doer, he just offered immigrants a legal way out of their jeopardy. We will have to consult the Congress on that, as a new law will have to be passed.

The last speaker of the day was Betty Klash, our general manager. "Dear members of this unique idea. Dear members of the press. I can't believe I'm here in New York. I am the person in charge of managing this incredible project. I just finished writing a book. It will soon be available. I might have to write another one because what you are hearing today will soon change the world and the way business is done, the way we travel, the way we think, and hopefully we will live to see the absence of conflict. We will live to see the collaboration between religions and the success of the democratic system in the world. I might add, the respect for equal rights for women, men, respect for different races, and their beliefs. I want you to feel the emotions that I feel, and if you do, it will be your obligation as the spokesmen and women for our incredible plan for the world, to transmit this to all the world. Starting with your communities and making sure that each one of your volunteers make this project viable and a reality. Tomorrow, we will make public our 18-point plan. This plan will surely be modified as time goes by. We will probably find new and better, or more important and more urgent matters to include. We will also eliminate the points as we manage and resolve them. I wish all of you success. Your success will be ours and mankind's. Thank You. May God be with all of you."

CHAPTER 16

THE SECOND DAY

That evening, many of our members got together in a hotel conference room to discuss the 18 points. Of this list, it was deemed that these are the most important and the ones we needed to resolve first: #1-Bad weather control. #3-World food and water supply. #5-disease control. #9-democracy for all. #12-equality in education. #15-nuclear control or its elimination. #16-world supreme court. #17-The world immigration problem. The rest of the points also need to be addressed but are not the top priority. I reiterated to the group that I understood their feelings and I was open to add additional points and eliminate points that might be duplicates of one or another. I emphasized that, in reality, it did not matter, because the contest participants would choose the portions of the contest that they wanted to resolve. The Judges would decide what or which of their result, findings or solutions were more appropriate, or realistic for the interests of the world. That is why the establishment of a World

Supreme Court was so important. It would be the ultimate judge of which was best, more convenient and doable, in all areas of conflict and decision making.

The conference began again exactly on time. Every one of the seats was occupied. The full house effect and the emotions of everyone present were clearly felt. The press, the cameras, the microphones were everywhere. Security personnel were visible in every corner. Fear was the only thing absent in this auditorium. The enthusiasm was so great that even the fear of anything happening seemed to be ignored. Most people knew that they were being watched and protected by the security details appropriately. Mike Frederick's opened the meeting and made comments on the opinions of some of the members as to presenting the most important points first. We have decided to present the 18 points together. The contestants will decide what points to select as their objectives. The afternoon session will deal with registration and application forms for the 18-point contest. We have opened a web site, WGNA18worldsolutions.com, and on that site anyone, whether a group, a company, an institution, or a lone individual can apply, and within 24 hours receive confirmation of their acceptance. Grants will be available for those in need for the initiation of their project. Every case will be studied accordingly. This afternoon, there will be a question and answer session. We hope to have sufficient know how to be able to answer all of your personal questions. We understand the restlessness at wishing to act faster and get things done more efficiently. Remember, this whole new concept started two weeks ago. The lottery win had really created a magical effect. Look how far and fast we have moved in such a short time.

Mr. Peers took the podium and reiterated that we need to calm down. We all noticed people were over excited about this incredible project. There will be time to resolve all the details that will develop. We know that there will be opposition to many of the points. Some of them can easily be criticized and they will be. We will accept all suggestions. We are open to everything from everyone. We also know that there will be opposition to our principals and to our way of presenting this and to our thinking in general, in many parts of the world. That is one of the beautiful things about a true democracy. Everyone is entitled to their opinion and can express them freely. One thing we know for sure, after this begins, there will be no force to stop it. I personally consider this to be the most important organization to start a world restructuring effort. A world we will really want to live in. We want harmony to prevail. We know that this contest will have the tools to make it happen. Strong applause followed. A bunch of the attendees got up chanting, "We want world peace." Another group got up shouting, "Stop nuclear proliferation." Another chanted, "We need to control the tornados, the hurricanes, the earthquakes, mud slides and uncontrollable wild fires, before they destroy the world."

I got up and began my talk. Dear members of this peace gathering. I was just informed of the presence of the Dali Lama in our auditorium. He is an active and revered humanitarian. His struggle for peace and freedom has made him one of the most-recognized and well-regarded spiritual leaders in the world. I am glad of his presence here in our spokesman conference. An explosive applause followed. He is one of many very important figures of modern society present. I know there are many United State senators and House of Representative members present. The Governors of several states asked for attendance entry, and we granted many foreign diplomats the possibility to attend. The ex-president of Mexico is here. Heavy applause began. I want to welcome all of you. Their presence gives even more validity to the establishment of this contest. They are here because they know that this is one of the most significant steps ever taken in modern times to create peace and understanding in the world. Every major newspaper in the world will carry all the information that is necessary to participate in our world contest, I am sure. Please, rest, have lunch, and come right back!

The flyers with the rules and the 18 points of the contest began to be distributed by our staff. Eighteen tables were set up, each one with sufficient staff to answer the specific questions that might come up for each point. The website had been prepared with the most logical and expected questions. The attendees started analyzing the points. The lunch vendors that had set up in the lower lobby were empty, few people bothered going down to eat. Everyone was too busy reading and commenting on the contest points. The website creators and staff were already receiving requests for application downloads from every corner of the world. By the end of the day over three thousand requests were downloaded. We intended to keep English as the official language but with the perfection of many apps like Google translate, the language factor was irrelevant. I was informed that that new apps have perfected the difficulties experienced in past translator applications.

The afternoon conference meeting did not commence on time. Everyone was too occupied with commenting and arguing about the different points and how to resolve each of them. An announcement had to be made on the loud speakers to the effect that the doors would close in 5 minutes, in order to resume without interruption, the last portion of the conference. That brought everyone to their seats and we began.

I got up to the podium and asked the audience for questions. To everyone's surprise, especially mine, every single one of the more than two thousand attendees got up and applauded my undertaking. The shouted congratulatory remarks and wished all of us, themselves included, lots of luck and success in this very important mission. To make this world friendlier, safer, and a better world for our children and grandchildren. They did not stop their clapping for what seemed an hour but must have been only 15 minutes. I was totally overwhelmed with emotion.

CHAPTER 17

THE WORLD REACTS

The next day, we received our daily newspaper outside our door as usual. I discovered that my picture was on the front page of the New York Times. Half of that page was about our 18-point contest. I also found our paid advertisement in the middle of the paper. I guess we could have dispensed with the advertisement. We turned on our Television. We were the headline in every news cast. We were the principal item in their program. *Fox* and *CNN* repeated our speech and our contest proposal every 30 minutes. Commentators talked about the implications of our contest and its possible results. One channel had a clairvoyant, assuring the host of our success. She had predicted, and been successful in 99% of recent happenings, or so she said. I checked our e-mails and we had received thousands of congratulatory letters from foreign country presidents, re-known investigators, university presidents, the Vatican and too many institutions and corporations to mention. A funny thing happened on the WGNA website.

Mike had opened a saving and checking account for our money activities and other necessary payments. The account ABA and account numbers were listed on the web site. There was a sudden deluge of donations from all over the world and to Mike's surprise a big portion of them came from China and India, the world most populous countries. We had never mentioned the word donation in our speeches nor interviews. What happened is that the common citizen, seeing a hopeful moment for future generations, decided to voluntarily contribute whatever they could to make this world better. We cannot make accurate interpretations of the events that began to occur, but our website was in the process of receiving millions of visits per minute, and Mike told me that the content was being translated into every language that Google had on the web. I was overwhelmed. My idea had started a world commotion. I had created a movement. The whole world was reacting. At 11 am, I received a call from the President of the United States congratulating me and offering us assistance in implementing our contest. I thanked him, and it occurred to me to ask him for a special visa waiver for the future contest delegations that would need to come to the United States to present and explain their ideas and discoveries. He told me that he would check that out with the State Department and with whatever Federal entity was necessary for its implementation. He gave me a special number to call if I needed his personal intervention in any contest related matter. Things were moving faster than a speeding rocket. Mike had joined Betty in our office and had taken full leave from any remaining bank responsibilities.

I called Betty and she told me that she just called in more secretarial temps. We needed extra help in handling the deluge of calls, messages, even telegrams were being received from countries with little or no internet service. She said that over a million application downloads had occurred since yesterday. The world's geniuses were organizing for the contest. I had created a world commotion. But a positive commotion, one where a positive future was envisioned.

CHAPTER 18

OUR FAMILY

Since the family meeting two weeks ago, many things had changed among our kids. They still called each other every day. They started to share their life events and made plans to meet frequently. They also saw that we meant when we said, "We want to be able to make a difference." They started getting calls from their friends and neighbors congratulating them on the recent activities generated by their parents. We had become celebrities. We transformed into doers and were news everywhere. Our children's lives had definitely changed. Our grandchildren who we rarely heard from, started to text us and call our cell phones using WhatsApp. They also had incredible life changes. They became super popular. Everyone wanted to meet them and socialize with them. The school principal and the university president gave them special attention and honor because of the activities their grandparents generated in the world. It seemed that everyone wanted to mingle with them in hopes of meeting their grandparents someday. Our children, who were very honest and were always very good kids, became concerned with all the activities we were getting involved in because they had difficulties reaching us, and when they did, asked us to take it easier and delegate some of our activities. They also wanted to know if they could be of help. They wanted to take part in our plan. Nelly, my wife, thought it was a good idea. I didn't want to get them involved in the beginning stage but thought they might be more useful later on. It was difficult to believe that we had reached the first of our objectives: to make our family aware that it was possible to make a difference so soon.

Our brothers and sisters and the rest of our families on both sides were also texting us and attempting to reach us. All of a sudden inviting us was above and beyond normal. They wanted us to get together with their kids and their friends, something that never occurred before. Everyone wanted to meet us and see if they could help or contribute in our project.

CHAPTER 19

APPLICATIONS GALORE

Forty-eight hours after our web site opened the public had downloaded over two million applications. This contest will be the first of its kind in many aspects. Today with the Internet revolution in full bloom there are very few things that cannot be performed with total exactness. Betty and Mike were getting ready to receive the completed applications and were preparing a team of scientist specializing in every area, to judge the reality and applicability of the proposal in every individual application. There would also be an economic analysis prepared with every proposal. We decided to economically assist every accepted proposal and make funds to work with available. Mike had informed that 150 billion dollars had already been received from the many new members that were coming in. Incredibly, there had been three billion in individual donations from common people who wanted this contest to succeed.

Betty had decided to hire a former Stanford University

president as the head of scientific recruitment. She sent us a list of professionals that we will definitely need for the 18 points. 1-petroleum engineers 2-acturial mathematics 3 -statisticians 4-chemists 5-reservoir engineers 6-geological engineers 7-production engineers 8- drilling engineers 9- nuclear engineers 10- industrial engineers 11-chemical engineers 12-biomedical engineers 13-electronic engineers 14-communication engineers , these include specialists in acoustics , defense , medical instruments , mobile phones, radio and satellite communication 15-computer science engineering 16-aerospace engineers, both astronautical and aerospace specialists 17-electrical engineers 18-material science engineers 19- physicists 20-statisticians 21-mechanical engineers 22-software engineers 23-business and information technology 24-economists 25-applied mathematician 26-civil engineers 27-architectural engineering 28-aviation management 29- biotechnologists. 30-operations and supply chain management 31-biochemists 32- geologists 33-security specialists 34- international business specialist 35-financial accountants 36-philosophers 37-molecular biologists 38-food scientists 39-investigative physicians 40- scientists in all areas of investigation. She sent this to the newly appointed man for immediate action. The immense quantities of applications expected was going to be a monumental task for this group of professionals. They also had to decide the grant amount of economic assistance each contestant group would need to complete and present their proposal. We decided to meet that same evening to decide and make a chronogram of events. I personally suggested that we put time limits on how long we needed to give the contestants to plan and send in their completed applications. They decided to give them a maximum of 21 days. 21 days should be sufficient for each of the contestants to gather the information and the necessary requirements to carry out their proposal and present it to the committee for approval or modification. The group that needs modification will have one week to attempt an acceptable one. After a lot of discussion, we decided that a maximum of three months was to be given to each approved contestant for each of the 18 points. If more time was needed, extensions would be given in as necessary, after they show the work performed or prepared up to that moment. After this meeting, Mike and Betty stayed on and we decided that we needed to implement rules in the meetings. Many of the active members were very wealthy and successful and were not used to meetings where they could not talk out of turn and interrupt every three minutes. We had to teach them that in these meetings, which were 100% voluntary, everyone had the same rights and privileges. I told Mike to have one of the lawyers of the group make a list of statues and procedure rules, so we could be more organized and civil in the next meeting. Believe me everyone was very excited. We had seen the immensity of the project, and the incredible effects it had on world opinion.

CHAPTER 20

THE RUSSIANS ATTACK

What we did not know is that there were others with different ideas. Mike received a phone call from one of the US government security agencies. He was asked to come into the Manhattan office for important security updates. Mike called me and told me of the call because it was a strange call. We had nothing to do with security or secret service agencies and we had no idea what to expect. Mike went to the meeting alone at ten in the morning. Ten minutes later he called me again, and asked me to please come to the meeting, it was very important. I was only 15 minutes away. I told Nelly where I was going and told her to get ready, that something new was developing. "I'll call you as soon as I know." Mike was very nervous again and he told me to listen to this. He let me hear a recording he had on his iPhone. The Russian Scientific society had decided to open a contest of their own. I was authorized to let you know what was going on in our security circles. We both went in for the complete briefing. We have been informed that the Russians are teaming up with the Chinese in an attempt to compete for quality contestants in the world. They are offering over five times the prize money for the winners of their contest. They are going to announce the contest in all the major newspapers. We thanked the government official and decided to call a meeting of the 12 most active and dedicated members for a 4 pm meeting in our office. Nelly and Betty prepared the room and called several important newspapers and the main networks and told them to have their reporters present for a 6 pm press conference.

I started the conversation after all the members were in their places. It is my opinion that their effort does not change our original plan. It is my opinion that what we should do is welcome them to our contest. This is not a political contest. This contest does not belong to the United States nor to any country. This is a world contest. Whatever their objectives are, it will add to our purpose. We don't care who resolves our world's problems first. We care about who resolves them effectively. I also think that increasing the prize money is no problem. We can modify this part and say that if more than one solution comes up we all win, all will receive the offered prize. The limit to the price will increase according to the importance of the solutions reached. Harvey Tower continued agreeing with my comments and telling us that he sees no problem in increasing the prizes tenfold if necessary. He considered the attempt to compete with us, as absurd and anti-scientific. Most of the other members agreed with my stance.

Then Mrs. Puig asked to speak. She said, "I would call them and ask them to join us as members. We have no intention to compete with them. In this type of world contest there is no competition between countries. All the countries participate. I mean the citizens of any country can participate, not the government of any country. I think if we make that clear to them, and if they are real scientists they will understand." Mike received a phone call and excused himself, it was important he said. He came back five minutes later and asked to speak. The Chinese foreign minister had just called our secretary of State and guaranteed him that they would not interfere with the contest. They would not join the Russian Scientific Contest. They would encourage any Chinese participation and participate economically with all positive outcomes. They agreed that the results will benefit China and the World. "It is time to create a cooperative world." The members applauded at the breaking news. We asked Mike to see if he could get the head of the Russian scientific group on the phone. He went out again and after inquiring, he was able to get the contact information. We decided to call with the speaker on. We introduced ourselves as the promoters of the world contest. The Russian spoke English very well and congratulated us on our extraordinary idea and effort. We then formally invited him and his board members to form part of our board of directors and join us. He was overwhelmed. He thanked us but said that the scientific institute was a government operated institution and he would need to receive permission to accept such an honorable position. He told us that he thought our offer was convenient to the present political feelings, and that we might receive a positive response. We thanked him, and he promised to call back as soon as possible. We decided to wait before making a declaration to the press.

It was almost 6 pm. All the outer and inner halls of the office were swarming with reporters. Betty went outside to explain that the meeting was taking longer than expected and that they would be briefed as soon as possible. At 7 pm *Izvestia*, the principal Russian newspaper, announced that The Russian Scientific Society had been invited to be part of the principal board of directors of the world contest. They congratulated Dr. Cantor and the members of the Association for such an intelligent decision. They said that the Russian government totally backed the contest. We had our answer and it was good. The reporters were called in and we announced the news from China and Russia. It was now going to be a real-world contest. Thanks God that was resolved. I was worried for a while.

CHAPTER 21

BETTY KLASH'S THOUGHTS

I am so extremely happy that Mike has called me to head this incredible contest as the general manager. In the beginning I thought that he had offered this to me because he wanted to be near me. Then I saw the reach that this contest was going to have, and the incredible international acceptance and I was overwhelmed both with the objectives and how fast and easy everything developed. It was great to work in an institution where money was no problem because there was more than could ever be needed, and the professionals and the members were the most educated and the richest on the planet. I have been going through a very uncomfortable divorce from a man that I married when I was young, and when money was something that I did not have. Marrying him took my family out of poverty. My father had lost his fortune after his factory burned down and the insurance group he was insured with folded. He was never indemnified. It was a marriage of convenience. He was older, very busy, and he wanted a beautiful smart Oxford University economist in his house. I was a showpiece for him. He never expressed or showed feelings for me. After 30 years, I decided that I wanted out. He said he would not permit it, and for the last 10 years he had been successful. Mike is the one I love. I met him at Oxford and we both fell in love. He went back to the United States and never asked me to be his wife. Then the guy I married offered me everything I needed at that moment in my life, and that's when I made the biggest mistake of my life. Mike and I want to marry as soon as I get the divorce. I will. I just hired a Swiss lawyer who said he found a loophole that will force him to accept the divorce decree. The project has me busy and happy and excited. To top it off, I see Mike day and night. He is simply fantastic. I want to adopt a child or two to make up for my childless years. I don't think I can still have children of my own. My future ex didn't want to have children. He never even tried. The other thing on my mind is the possibility of terrorist involvement. All Europeans fear that. The fact that many of the points in our contest directly and indirectly offend these people's beliefs. They might attempt to boycott our contest. We have to be alert and I will suggest that our members keep this in mind, in every phase and in every place where we meet. We must warn all the participants of the inherent dangers involved.

CHAPTER 22

OUR FIRST APPLICANTS

Today we received applications and the name on the first application was Arshia Bhatia. His objective is Point #13. He will develop a foolproof world electoral system. He is a Magna Cum Laude graduate of Hi Tech Institute of Engineering and Technology (HIET) in Delhi-Hapur, India. He also attended Georgia Tech in Atlanta, USA, for his post graduate training. He applied with a team of six and said that he needed approximately $400 thousand for his experimental development of the impossible-to-alter machine. He added that he initially needed only $20 thousand to buy and make the necessary components. He added that he would need the rest or part of the rest to make sufficient sample models to test in different parts of the world. He said that he was ready to begin immediately.

Then we received three applications for Point #8 for appetite control. One from the United States, headed by an endocrinologist Dr Solo Hamilton, a Harvard graduate. He said he did not need any money for his project to begin. An Australian, Dr Scott Terri, from The University of South Wales, Australia, asked for $18 thousand for expenses and research and said that he had been experimenting successfully with a still unpublished miracle drug. The third applicant was from Singapore, her name was Shilla Taso, and she was a medical laboratory technician graduated from the University of Cincinnati in the United States. She had been experimenting with rabbits and had gotten positive results. She did not need money to begin her human research project. She understood that she only had three months to prove her experiment worked in humans also. "It is too short a period, but I will prove the success of my experiment."

We received one application from a group of lawyers in Switzerland attempting to get approval for Point #16. They would interview Nobel Prize winners and the most notable jurists in the democratic world and present a list of ready and willing judges for The Supreme World Court. They considered themselves ideal for the selection process especially because the Swiss have always been neutral and would not select Judges with negative influences in world matters. They were represented by a distinguished International law specialist, Professor Herman Mayer.

Two applications for Weather Control Point #1 arrived, one from a team of Chinese and Japanese investigators, headed by Dr. Shinto Umami, and the other from the University of Tokyo. They have been working on a solution for neutralizing hurricanes, typhoons, and cyclones.

They are the same storms, the differences in names has to do with where they occur. Hurricane is used in the North Atlantic. Typhoon in the Southern Hemisphere and Indian Ocean. They think that their investigations would also apply to Tornadoes. They limited the future results to storms that produce wind. They made a point in not including, floods and earthquakes. They asked for $5 million plus the use of an airplane capable of flying into storms.

Seven applications focused on Point #7, a system to shorten travel time. The teams were from different countries. The USA, The Netherlands, South Korea, Israel, Italy, China and Denmark. They all had different plans for different modes of transportation and each solicited several million dollars to begin their experiments. They all had incredible ideas. The question each one posed is that there was a difference between theory and practice. All seven complained of the very short three-month period for the contest.

A team from Germany applied for Point #2, traffic control. They said that they have been working on several projects for over three years and that in the three-month period they would be able to make several prototype models. They needed $42 million upfront.

We had several groups apply for Point #3 for resolving the food and clean water problem for the growing world population. The groups were from Israel, Argentina, Thailand, Iceland, Vietnam, Finland, Burma, US, Nigeria, and Canada. Each one had a different approach. Iceland and Finland concentrated on fish farming. Thailand and Burma and Vietnam on the mass production of rice. The United States and Canada and Argentina in wheat production and the US and Argentina and Uruguay in cattle farming. Israel focused on all agricultural products, and one of their many specialties was conversion of salt water into clean drinkable water a very important world problem.

The Swiss, French, Japanese, Chinese and German and 12 other countries had several applicants that applied for Point #12. The System for Universal education. We expected an incredible number of applications for this area. There were many points of view in this area and it was going to be very difficult in choosing an ideal system. We knew that we would probably have to select a combination of the best of several applicants' proposals.

As for Point #15 the Nuclear Weapon control or abolishment point, we were surprised that Russia, USA, Japan, and Pakistan applicants registered first. The asked for more time and for different amounts as working capital. Only one spoke of their mode of operation. All the applicants were headed by very well-known nuclear physicist and nuclear engineers. One of the groups asked for $125 million. They said they were working on a vibration ray that neutralized all nuclear armament.

Point #5 had more than 40 applicants representing 8 countries. Israel alone had 22 applicants. In terms of battling disease, cancer, heart disease, infectious and hereditary diseases, were the principal areas of concern. Almost every applicant had a specific plan for various illnesses. Cancer and

HIV were the main issues covered. The economic needed for each one was varied. several of the applicants wanted to compete in the contest and asked for no economic cooperation.

Today, only a few days after all this began, we knew that additional time was needed for the groups to gather, organize and meet to implement ideas in their area of expertise. Of course, we were going to give them the necessary time they required. Our project had become a mega project. After a week, over two and a half million applications had been downloaded. I asked myself, "What have I created, will we be able to manage this?" Those questions lingered in my mind, day and night. After all what could go wrong? I answered myself, nothing at all. I knew that this positive and incredible project would bring incredible results in all areas and for all the citizens of the world."

CHAPTER 23

THE MEMBERS MEET TO RE-AFFIRM GOALS

Today was the 10th day after the beginning of our information blitz, and everything was running smoothly. We were constantly receiving questions from our members. The members could be classified as the very active, active, and donors only. We had over 1800 members, and many wanted constant and continuous updates. So, we sent out 1800 emails and prepared a 3000-seat auditorium for an update meeting. We asked for a R.S.V.P. so that we could properly attend to all participants. Some answered immediately, and we also knew that many would show up without an acknowledgement. We knew they would bring family or friends, so we had to be prepared for surprises. These members had a right to be informed. Some of them had invested hundreds of millions of dollars, and they were really keen to receive prospective results. Many felt that they were the pioneers in our contest and would be registered as men and women who made the world become more human, friendlier, and maybe even war free. Those were our real goals. This made it their goals. Their money and participation were going to make this possible.

To our surprise we filled the auditorium completely with members and many additional family companions and. of course, the news media were present. We really did not invite them, but it was difficult to keep the information away from them. Our project was the news of the century. We never even tried to avoid them. All publicity obtained was positive for our contest.

The purpose of our reunion was to give an update with all the information we had and answer questions and, of course, accept suggestions and advice. During the conversation and explanation, a man got up and insisted we make the donations to the contest public. One of the contractual points in the acceptance of donations, or as we call them, "investments in making a better world," was to keep the identities of the different "investment" owners private from everyone except the IRS in the United States. We also explained that many of the "investments" came from foreign donors. In many countries these types of donations are not seen as good signs, and when discovered may lead to kidnapping and or harassment of the donating families. We reminded the audience that there were too many countries involved with undemocratic governing systems. This is where real freedom was questionable. Another question came up to the importance of us showing what we were spending, since we had already informed the members of our vast money availability. We decided to keep that information as private as possible because publicizing it would stimulate future new participants in increasing unnecessarily the amounts they solicit, instead of the amounts they really need for the ideal finalization of their project. Someone asked questions relating to security and anti-terrorism procedures, that is, they wanted to know what we were implementing for our safety. Again, we explained that we were receiving advice from many world institutions and we didn't think it would be too smart to let these people with murderous intentions be forewarned as to what we intend to do or to what we have already done to catch them or to impede their unacceptable actions. Then surprisingly, a member asked us if it was true that the Russians were attempting to create a similar contest of their own. This time I decided to answer. "Dear friends, I want you to know that three of the prestigious Russian Institute of Scientific Investigations have applied and were accepted by our Board. They will collaborate with our scientists and I want you to know that this morning 22 applications were received from several Russian cities for different points." Then a new Chinese member got up to confirm this fact, and to add, that one of the main institutes in China had joined a Russian group from Vladivostok in the development of a love app, Point #11. Everyone laughed at the implications. Mr. Kitza, our Japanese member of the Board, announced that a Japanese firm had just applied for the contest for Point #8 appetite control. They had two innovative systems that would change the worlds eating habits.

A very fat relative of one of the attendees got up and thanked God for creating this contest, "To give me my life back." Several other overweight men and women got up to applaud his remarks. A lady that had a relative who was battling cancer for a decade got up and asked that Point #5 be promoted as much as possible. "I want my uncle to be able to use the results. I want him to be able to be in time to treat his cancer effectively." Another lady accompanying one of our members got up to make the observation that she had a nephew who was a technology genius and that he had developed a machine that permits voters to vote and that it was fantastic. The results would be immediate and unalterable, guaranteed. I suggested that he download an application and join the contest. She said that it would be difficult because he was only 12 years old. I told her that age was not an obstacle. A very old man in his nineties got up to ask why they didn't add a point to prolong life beyond the statistical average. Mike said that he would suggest it to the board. Mike then went on to say that if we control the major reasons for early death and people live and eat healthy, people will be able to live longer without needing to integrate that point in the contest. Then came the question as to who would judge the contest and who would implement the results? Would we use the fund's money to make the results viable? Could more than one contestant win per point? How can the judges be impartial? We decided to answer that no one should worry. The future judges will be selected from the cream of the cream in the world, they will be just and impartial. The rest of the answers to all your question will be answered as time passes. We will make the right decisions and you will be informed and be a decisive part of our actions.

The meeting was successful, no questions remained unanswered, everyone left with a satisfactory grin in their face. If I had a mechanism of weighing satisfaction, I knew it would reach 99.9% out of 100%. I could feel it in the air. I could feel it in my bones.

CHAPTER 24

SLEEP DEPRIVATION (POINT #4)

I asked myself why I included this point in the contest, and I guess it was because too many of the people my age continuously talk about their inability to fall and stay asleep for an adequate time. There are many reasons that people get inadequate sleep and become sleep deprived, from individual demands at work and even home life and to sleep problems like insomnia. Depending on the degree of sleep deprivation—both how little we sleep and for how long we are sleep deprived—there can begin to be important consequences to our health and well-being. Total sleep deprivation, in which no sleep is obtained for several nights in a row, certainly can be a trigger to create the sleeping problem. Continually obtaining too few hours of rest may likewise have a cumulative effect. The degree of sleep deprivation required to start to experience side effects likely varies for each person depending on their individual sleep needs and genetic predisposition.

If someone needs 10 hours of sleep to feel rested, but only gets 8 hours, they will gradually become sleep deprived even though they may seem to be getting enough sleep based on the population average. Sleep deprivation can affect both children and adults. Adolescents with delayed sleep phase syndrome may have difficulty meeting their sleep needs due to a delay in the onset of sleep and required wake times for school. Also, there are consequences that might have to do with sleep deprivation or not sleeping enough. Hallucinations is one of them, here the perception of something that is not present in the environment, as opposed to an illusion, which is the misinterpretation of something that is present. For example, seeing a dog where there is nothing is a hallucination, but mistaking your coat rack for a person is an illusion. Depending on the length of sleep deprivation, approximately 80% of normal people in the population will eventually hallucinate. Most of these are visual hallucinations. In contrast, people with schizophrenia often have auditory hallucinations, hearing sounds (often voices) that are not there. Sleep deprivation can cause other symptoms that mimic mental illness, such as disorientation and paranoid thoughts. The affected person may be confused about details related to time or location. Paranoia may lead to feelings of persecution. In fact, one study found that 2% of 350 people who were sleep deprived for 112 hours experienced temporary conditions that were similar to acute paranoid schizophrenia. These are the principal reasons why I thought this subject was important. When a person does not get sufficient sleep time, there is an inability to have a normal working and thinking day. That is the main reason for Point #4.

CHAPTER 25

A SYSTEM THAT SHORTENS TRAVEL TIME (POINT #7)

A few months ago, my wife and I went on a weekend trip from Miami to Panama City Florida. During the road trip we were surprised how long a trip it was, many times we were limited to areas that had reduced speeds and it seemed to take us forever to get to our objective. before deciding to go via highways we had googled possibilities of flying to Panama City and saw that the only way of getting there from our area airports, was going either Charleston, Virginia or Atlanta, Georgia. At a fast glance at the possible connections, and wait time at airports, we concluded they would take even longer than driving. I then thought that there was new news of the Hyperloop, which will reinvent transportation and maybe eliminate the barriers of distance and time. The name "hyperloop" is a trademark and is registered to SpaceX. According to Elon Musk, it is a system to build a tube over or under the ground. High speed fans would compress and push the transport media to its destinations at incredible speeds of up to 760 miles per hour, that's approximately the speed of sound. Musk envisions this as being powered by solar energy. "it would be able to withstand extreme weather and earthquakes. He goes on to illustrate that a trip from Los Angeles to San Francisco, which takes 6 hours would diminish the travel time to 35 minutes. That would really revolutionize transportation. Can you imagine living near the San Francisco bridge and traveling to your office in NYC in a regular commute of a few hours? This information made me place Point #7 as one of the 18 in the contest. There was a Hyper pod Competition in the early 2016, with 115 finalists selected from over 1,200 submissions from around the world. Thirty teams with the most promising concepts were asked to build prototypes and invited to compete in California in January 2017. Of the 30, 3 were unable to meet deadline and after preliminary testing only 3 were deemed suitable and allowed to run the 1-mile Hyperloop test track built by SpaceX. Deft University of Technology in the Netherlands received the highest overall score. Several participating universities were mentioned for their safety and reliability. MIT and The Technical University of Munich were the only one to complete the complete loop. There are many private efforts underway to test in new sites. Recently a 1.8-mile track was built in North Las Vegas and a firm HTT is forging plans to build a network in Europe connecting, Vienna, Bratislava, and Budapest. All these efforts will likely result in supersonic travel. Our Contest will definitely promote this idea, and others in the realm of air, shipping and space travel.

Sir Richard Branson, the founder of the Virgin Group, has an eye toward future supersonic flights. He predicts that "people will be able to travel around the world in no time" and in a "friendly and in an environmentally safe way." Branson's Virgin Galactic has partnered with startup Boom Technology Inc. to work on jets with passengers with a flight time of 3.5 hours between New York City and London. That is 2.6 times faster than today's planes. According to the same report one of their models is scheduled to be tried out in 2018 and if all goes well, as expected, they will begin passenger service by 2020.

CHAPTER 26

THE WORLD'S IMMIGRATION PROBLEM (POINT #17)

In Point #17, the immigration facts are plain to see and detect according to the United Nation. It is estimated that 46.6 million people living in the United States were not born there. The US immigrant population is nearly four times that of the world's next largest immigrant destination, which is Germany, with about 12 million immigrants. A United Nation report said in 2005 that there were 244-million international immigrants worldwide. We all know this has increased dramatically in the last few years.

International migration is a global phenomenon that is growing in scope, complexity and impact. Most of the immigrants attempt to settle in a country that gives them hope of survival, jobs, education, and freedom. They try to escape from countries in war, political strife, after floods, hurricanes and other environmental disasters. Some immigrate because of their religion others because they are persecuted where they live, for example from the recent wars in Iraq, Syria, and Afghanistan. Strife and hunger in Africa have increased immigration. Most, of the immigration has been illegal. Destination countries have not been prepared for the huge deluge of immigrants. Most of these immigrants come without any economic possibilities and without any know-how as to the way the outside world works and conducts itself. This creates technical and economic problems for the countries that they choose. In recent years the political and economic absurdities happening in countries like Venezuela have increased this illegal immigration tenfold. Their next-door neighbors have suffered due to these arrivals and lack of resources required to cover the needs of these refugees for food, shelter and medical assistance.

I know that there are many countries in the world that have a small population and these countries would like to increase it. Some European countries have a very low birth rate. They seek to attract new immigrants by selecting them according to their necessities and are very strict in allowing non-essential types of immigration. My idea is that we must improve life in general in the world by providing mechanisms where all countries resolve their internal food, employment, housing, health, education and operational methods. Immigration will diminish everywhere. This is, of course, a long-term situation. There are countries that are overpopulated and have the need to ease their outward migration to reach their goals. These countries can offer countries in need of migrants the ability to get them selectively. My aim in this point is to create an equitable distribution of population without affecting the operation and functionality of the immigrant-acquiring country.

CHAPTER 27

NUCLEAR ENERGY FOR POSITIVE USE (POINT #15)

Since the beginnings of 1950, when the first nuclear power station to produce electricity was implemented, electricity was produced from the heat produced of the splitting of the uranium atom. Electricity then became a cheap, clean system of producing energy in many advanced countries. Radioisotopes were developed as early as 1911 by Georges de Hevesy.e went on to win a Noble prize He went on t He went on to win the Nobel Prize in 1943 and then again in 1959. He was the first to use radioactive tracers, now these are of routine use in environmental science. Isotopes are used in agriculture. They are used in plant mutation breeding, in fertilizers, insect control, in consumer products, in food irradiation, in industry, in inspection and instrumentation, in carbon dating in desalinization, in medicine, transport in water resource and the environment. In other words, nuclear energy has many positive applications and is not limited to the implementation of bombs that can cause mass destruction. The bombs are the elements that have to be put under a strict control. This of course is very difficult if not impossible until we can neutralize the dictatorial governments that have them and with which the express their supremacy over others. The fear of dealing with this type of government has destabilized many countries who are in fear of antagonizing their ruler and decision maker. One day after we achieve our goal and have these countries within a real democratic system, we can have the future New World Court, Point #16, outlaw its use and production as well as continuously stockpiling them. All these bombs will need to be deactivated and made useless and buried in the deepest corner of the world. Only this way will our world survive from the atrocious words and accusations that these leaders throw at each other. If we can implement contest Point #15 and make it work, we will have succeeded in taking a gigantic a step to world peace. This is our goal. I know we can do it.

CHAPTER 28

APPETITE CONTROL (POINT #8)

Everyone says that its easier said than done, I personally thought that most people looked at overweight people as disorganized, over stressed, or depressed, who think that by satisfying their hunger pangs, their problems would either go away or diminish or disappear. They were wishful thinking men and woman who did not see the reality of life and their future looks, health, and future job search ability ever being affected. They did not see the harassing and bulling that their overweight would generate. Eating seemed to make them happy, and jolly. There is belief somewhere, that fat people are nice, friendlier, people, that do not get old fast or that look younger even though they are older. The truth to the matter is that overweight is a problem that can really cause severe health issues and an early death. It is easily noted that some places have an excess of overweight people. The reasons differ. Some have a genetic predisposition to gaining weight easily and others because the food intake habits are very high in caloric products. Then there are those who are overweight due to hormonal disbalances, thyroid problems and other health issues. I personally have been dealing with a weight issue, and whenever I finally reach a goal, I seem to regain the weight in no time. My successful weight control friends say that it is a matter of discipline. Well I decided to place Point #7 in the contest because I know a great portion of the world lives with this problem. Let's help them find a solution.

CHAPTER 29

FOOD FOR THE WORLD (POINT #3)

The objective of this portion of the contest Point #3 is to find a remedy to the world hunger problem. We want a solution so that food is not an obstacle stopping humans from reaching their potential. We want to create a situation where farmers can grow sufficient food to feed the world, and where children can dedicate themselves to studying and not to spend a big portion of their day searching for food. Where nations have the possibilities of responding and helping others in time of crisis and natural disasters. I think that the contestants will bring in ideas that will begin to resolve this acute world problem. Eliminating hunger, and by doing that, eliminating malnutrition is among the most intractable problem humanity faces today. High prices continue to be a factor. To add to this, the mainly affected people are those stricken by poverty. Hunger and poverty go hand in hand. Even though there might be sufficient food in particular markets, poor people's inadequate income creates a situation where they cannot afford to purchase it. They are too poor to afford it.

Another acute problem that we cannot ignore is water scarcity or lack of sufficient available fresh water resources to meet the demand. This problem affects every continent and is considered as one of the biggest global risk factors to resolve over the next decade. According to the World Economic Forum, four billion people live under conditions of severe water scarcity all year. These include the fact that half the world's largest cities experience frequent water scarcity. We need to know the facts. "Although a mere 0.014% of all water on Earth is both fresh and easily accessible of the remaining water, 97% is saline and much of the remaining 3% is hard to access. Technically, there is a sufficient amount of freshwater on a global scale, for humanity to get by. However, due to unequal distribution, resulting in some very wet and some very dry geographic locations, plus a sharp rise in global freshwater demand in recent decades, humanity is facing a water crisis, with demand expected to outstrip supply by 40% in 2030, if current trends continue. The essence of global water scarcity is the geographic and temporal mismatch between freshwater demand and availability. The increasing world population, improving living standards, changing consumption patterns, and expansion of irrigated agriculture are the main driving forces for the rising global demand for water. Climate change, such as altered weather-patterns (including droughts or floods), deforestation, increased pollution, and wasteful use of water can cause insufficient supply. At the global level and on an annual basis, enough freshwater is available to meet such demand, but spatial and temporal variations of water demand and availability are large, leading to (physical) water scarcity in several parts of the world during specific times of the year. All causes of water scarcity are related to human interference with the water cycle. Scarcity varies over time as a result of natural hydrological variability, but varies even more so as a function of prevailing economic policy, planning and management approaches. Scarcity can be expected to intensify with most forms of economic development, but, if correctly identified, many of its causes can be predicted, avoided or mitigated.

Some countries have already proven that decoupling water use from economic growth is possible. For example, in Australia, water consumption declined by 40% between 2001 and 2009 while the economy grew by more than 30%. The International Resource Panel of the UN states that governments have tended to invest heavily in largely inefficient solutions, mega-projects like dams, canals, aqueducts, pipelines and water reservoirs, which are generally neither environmentally sustainable nor economically viable. The most cost-effective way of decoupling water use from economic growth, according to the scientific panel, is for governments to create holistic water management plans that take into account the entire water cycle, from source to distribution, economic use, treatment, recycling, reuse and return to the environment.

It is our hope and goal that the contest will bring new ideas on how to resolve this very important world problem. We look forward to feedback from Israeli technology companies that have made miracles in this area. They have created a paradise from a desert. The Israelis have developed ways of converting sea water into useable and totally safe drinking and irrigating water. We will achieve our goal, here also.

CHAPTER 30

A COMPREHENSIVE SYSTEM TO TEST COMPATIBILITY BETWEEN COUPLES (POINT #11)

This point was initially criticized by some of the members of our group. One of the men asked me if it was a joke. Another asked me if I was having personal marital problems. I will try to explain my reasons for this point. Too many of today's couples get divorced, some after only a few years together. At times, I feel that there is either a crisis or an epidemic in divorce courts. The problem is that these divorces cause many unsolvable problems, especially in the matter of children, which are very affected. Children of divorced parents usually have handicaps of many types. These handicaps usually create secondary and tertiary problems in their lives. With our contest, I think that one or more solutions will be brought to the world to look at, to investigate, and see if all the inconveniences that occur, can be avoided in the already married, and for the future generations that plan to marry. Maybe we can develop a way to choose our future mates with sufficient statistical and personal data so as to be able to predict with little or no margin of error the possibilities of future conflict. With today and tomorrow's technological findings, I feel confident we are closer than ever to this being realistic. Can you imagine a world where couples have genetic predisposition to forgive and forget? This point will definitely contribute to world peace.

CHAPTER 31

HAIR GROWTH (POINT #6)

This is another point, of the relatively minor importance in this contest. The fact is, that it is important for some people and in today's world of appearances, it is even more important. Many men who lose their hair at an early age tend to have some type inferiority complex. Some believe that this may affect their possibilities of getting their ideal job or their ideal mate. Many women have a similar problem, especially at an advanced age, where hair thinning and hair loss happens more frequently, and where a wig is not the solution. It's true this is not really important, but some feel hair is their most precious body part. They tend to do incredible things to avoid losing it. Incredible investments are made to make it look more beautiful. It's just a small point that may have a solution. If a solution is found, I am sure it will make many millions of the world's habitants happy, and this, will also contribute with a small grain of sand, to a better and more harmonious world.

CHAPTER 32

EDUCATION FOR AL (POINT #12)

This point is really a dream. Wouldn't it be great if children in Africa could receive the same education as the children in Canada, Switzerland, Israel, and the USA? It would create a world with better equilibrium, a world where everyone has the same opportunity to achieve goals. It would be a gigantic step for the countries that are ridden with poverty, hunger, famine. Can one imagine how incredible the world would be? Wars would probably be a thing of the past. They would disappear. When everyone receives the same education, has the same aptitudes, and opportunities and uses similar knowledge data, wars are one thing that they would want to avoid. It would interfere with their idea of progress, their idea of making their life and their future generations better. Of course, this is a simplification of the reality, and of the time and effort that it will take to achieve such an incredible but possible goal. If we all gather with sincere enthusiasm, we can make it a reality. Of course, it is not the citizens who look for war it's their leaders.

CHAPTER 33

THE CONTEST ADVANCES

The contest was receiving world attention. Every system of mass communication was on top of it. Rare was the media source not constantly reporting on our progress. Our office was deluged with hundreds of thousands of questions. Explanatory statements were constantly and continuously given both orally and in writing. Our web site had to be continually updated and redesigned. The contest points #9, 10, 14, 15 and 18 lacked applicants. Presumably, because of the nature of the points and the obvious impossibility of creating a realistic organization to create and regulate such difficult organizations. Point #9 refers to a military force. Point #10, For a world body to substitute the not functioning United Nations. Point #14, Is all about space exploration and nations cooperation and union. Point #15, Elimination of the nuclear arsenal. Point #18, the very difficult to resolve, Israeli-Palestinian conflict. My hope was that as soon as Point #16, the implementation of a Supreme World Court, became a reality, all the difficult points that I just mentioned would be addressed and become reality. Its mere presence and acceptance would give authority to all the world organizations that were needed.

We were getting to a stage where everything was running on automatic. We had to extend the three-month initial period to five months and promised continued updates according to the particular needs of the different contestants. 90% of the applications had received preliminary acceptance. As soon as the special committee gave them total approval, an economic committee studied their financial needs and funds were sent to them. Each contestant had to send in a weekly report as to the use of the funds, with the corresponding invoice as proof of its valid use. To our amazement, some countries that we would have never thought possible, participated. An example is Madagascar, Tunisia, Andorra, Lichtenstein, and San Marino. The countries that had the most participants were of course the largest and most populated. India, China, Russia, USA, Canada, Brazil, Mexico, Australia, and an exception to size and population was Israel with as many participants as Russia and China. The Points with the most participants where Points #2,3,5, and 12. There were many people who thought they could resolve the better travel control app. Create better world food production, Cure Cancer and many other diseases. Resolve the unification, quality, and standardization of education in the world. Things were moving along fast. My motivation was stronger than ever. We had caused a world revolution with my idea of this contest. I felt like I had already achieved my initial quest to make a difference.

CHAPTER 34

THE NOBEL PRIZE

As everyone knows the Nobel Prize is a set of international awards given by Swedish and Norwegian institutions in recognition of outstanding academic or scientific advances and research. This prize began in 1901, when Alfred Nobel instituted this very successful prize and it continues today. We planned to contact the institution to ask them to collaborate with us when one of one of the distinguished members of our board contacted us and explained the possibility of a conflict with the institution. I was told that some of the deciding members thought we were stepping into their fields and saw us as a possible conflict of interest party. We decided to send an emissary to meet with their board to explain our mission. In reality, we were looking for many possible Nobel Prize candidates and at no time did we want to compete with such a distinguished society. The fact is that we were generating so much positive publicity and the expectations of our results were being anticipated with such enthusiasm that some of the Nobel Committee associates were feeling interference in their future Noble Prize announcements. I decided to choose a multilingual envoy and one who had charisma and was used to dealing with diplomatic-circle individuals. He flew to Stockholm and reported back the great difficulty he had in getting an appointment with the proper committee spokesmen. After hearing that, we all decided that it was the moment to take up the White House offer to help us with any problem, we might encounter. The incredible thing is that merely 10 minutes after our call to the special number the US president had supplied me with, we received a note from the Noble Committee that they would immediately receive our emissary. The meeting went well, and the problem was resolved in no time, with no complications or conditions. Our representative made it clear to the Nobel Organization that the results of the incredible contest we were promoting would give them even more possible candidates for their annual selection. This clarification was greeted with a standing ovation. I later received a letter placing me on the list for The Nobel Peace prize nominees for this year. The nomination was made by the Nobel committee themselves. I thanked them for the honor and insisted that there were many other worthier Peace Prize nominees available for nomination.

CHAPTER 35

MY THOUGHTS

I sat down one morning and tried to make a resume of all the accomplishments we had achieved. We had really converted an idea into a global affair. It was miraculous. I asked myself what had happened? Was it the effects of me winning the lottery? Was this a magical effect of winning the lottery? Was it the power of having all that money suddenly? Because to be very honest, I still had the money I won. Little if any was necessary to use for my contest project. What happened is that I became a world figure because of my contest proposal. The only ones that know of my winning the lottery are my immediate family and my new associates, Mike Fredericks and Betty Klash. The rest of world have no idea, or that's what I am led to believe. Everyone gave me their word to keep everything secret for the benefit of all involved. We did not want to be deluged by charity seekers and predators of all kinds. I got to where I am because of the influence that Mike Fredericks gave me by introducing me to the right people and proceeding to establish this incredible, or should I say fantastic, organization. The investments, or should I say donations, that have poured into our bank accounts are so vast and immense that we could probably resolve many of the health, education, infrastructure of most third world countries in the world. Those funds are ready to be spent only on the 18 points of our contest. We will worry if we need more, or we will worry what to do with whatever is left over after we complete our objectives. This venture has been very difficult on Nelly and me. We have dedicated every available second to our plan. Our children and grandchildren have complained of our absence from the family scene. They in turn have become famous. The fame we obtained through our idea to make a world contest. We have been able to implement our idea because of the fame we have achieved or because of it.

Nelly and I discussed our next move. Should we leave the continuation of the project to the already formed and functioning organization and just hang out as cautious and ever-present detectives so that everything is carried out properly? Or should we continue to be on top of every little detail as the project gets bigger and bigger as every moment passes. We are not as young and energetic as we used to be. Will all the stress that this causes affect our health? Nelly's answer was immediate. My dear Dr Cantor, I think that what you just said is premature. The organization needs you. Everyone sees you as the promotor, the head, the leader of this incredible idea. You cannot stop. There is no one who can take your place at this time. Maybe after the six-month period ends?

After all the preliminary results come in. That's when you can start delegating your leadership to the different department heads that we will need to help designate. Then there is the question of the World Court. The court will need a leader and spokesman, and you will be the natural person to lead it. Nelly, that is one hell of a responsibility. That's a lifelong responsibility, do you want me to take that upon my head? I know of no one more fair and honest and whose leadership cannot be in doubt than you.

CHAPTER 36

NOT EVERYONE LIKES WHAT WE ARE DOING

What was happening in the world was very exciting for most of the people in the free world. The thing that most of these men and woman of the world did not know or did not think about, is that the results of this incredible contest were a reason of extreme worry. There would be definite negative results on their way of living, their way of ruling their country. If the results of some of the points were brought to the world limelight and made known to everyone, these men and women would eventually lose the power they had in their countries and in the world scenario.

The rogue nations, the ones governed by life dictators were the one group that were very concerned with the outcome of our quest for real democratic principles. They saw that some of the points would strive to eliminate them from power. The world would definitely change with all the contest results. Another group that would be affected would be the pharmaceutical companies everywhere, especially the ones that produced products that cure the illnesses that the contest would resolve. Once these illnesses were eliminated a great part of their business would disappear. There were other groups that were concerned with the outcome of the contest. Groups that would be affected by the substitution of methods of transportation, groups involved in food production, groups that benefitted from the results of the catastrophes that bad weather produce and groups that were active in terrorist activities, both due to political change and to extreme religious beliefs. These, and many more, were potential enemies of the contest idea because of the negative results it would cause to their governments and their world corporations and to certain other mischievous organizations.

CHAPTER 37

THE DICTATORS OF THE WORLD CALL FOR AN URGENT MEETING

It all started with a trip by one of the South American country's dictators. This trip was another attempt to obtain a loan from one of the big oil-producing nations of the world. The country he ruled, although one of the richest countries in the world, had depleted their monetary reserves because of vast corruption of its officials and his government cronies. The people of his country now lacked food, drugs, and everything else. The country's infrastructure was in ruins due to the complete absence of upkeep of their oil industry. The dictator needed fresh money urgently and he was willing to mortgage his remaining mineral reserves for another multimillion long-term loan. Many of its citizens were fleeing across the borders to their neighboring countries. The contest that had been proposed was really something that he feared. He knew that the consequences of the outcome would be the thing that would definitely oust him from power. If the world unites and eliminates all non-democratic countries, "I am lost," he thought. He sent out messages to his allies, to all the countries that had loaned money to his government, to all the presidents that ruled their countries undemocratically and disrespected human rights, and the right to democratically elect their rulers. His message was clear: unite against the results of this new phenomenon. This contest was aimed at their destruction.

Many of these rulers were extremely concerned with their probable loss of power in the very near future and thought that the idea of an urgent meeting was a good idea. After discussing several possible countries where the meeting could take place, the Cuban dictator urged all of them not to waste more time. "All my allies are always welcome to come to La Habana." Due to the extreme importance of this reunion, the Cuban government set up a date and suggested that the meeting should be for Presidents, exclusively. He also suggested that this very important meeting should occur with the highest degree of secrecy. If the world press finds out, we will not be able to carry out our counter measures effectively. The suggestions were difficult to keep secret in this modern high-tech world. As everyone knows, there are hidden microphones and micro cameras everywhere. At times, it seemed that even the tree branches overhear our conversations, and some believed that many of these new gadgets could be implanted in the body and transmitted even thoughts.

The following countries immediately answered: Iran, North Korea, Venezuela, Bolivia, Nicaragua and several small Caribbean islands and several African and Asian countries and Cuba of course. The organizers, Venezuela and Cuba, were very surprised that Russia and China were not interested in participating. They considered themselves unthreatened by the contest. In fact, China Television presented an editorial calling the contest, "The most innovative and interesting proposal of the century." To add to that, both countries had an increasing number of participants, all encouraged by government officials.

One interesting detail was suggested by the Bolivian President. We will finally get to meet the supreme leader of North Korea. He has only left his country to go to China and to Singapore, where he met with the US president. He was afraid of being assassinated. This will be an interesting meeting, he thought.

CHAPTER 38

THE WORLD'S SECURITY ORGANIZATIONS ARE ALERTED

I t all started with an email. Interpol (international police) had received an unidentified message from an unknown source to the effect that The North Korean government had requested permission to fly over Russia and the rest of Europe with Cuba as its destination. At the same time a similar request had been sent by Iran's government for permission to fly on the same date. Many more authorization requests were received from the different anti-democratic countries.

All this information was picked up by the security agencies of the free world. It was not a secret meeting any longer. The *New York Times* made mention of it in its early edition the same day. *CNN* and *Fox*, and all the major news outlets made comments about the coincidences of all the same permission requests. The fact that it became news immediately became a problem for all the attendees to the conference. The element of secrecy was no longer there.

The head of the security division of the North Atlantic Treaty Organization (NATO) alerted individual member countries of the implication of such a dangerous meeting.

The security heads of the United States, England, France and Italy immediately convened an urgent conference. All surveillance satellites were programmed to follow all movements in and out of the countries in question.

CHAPTER 39

A SECRET WORLD SECURITY ORGANIZATION MAKES A PLAN AND GETS READY

In the aftermath of the September 11 terrorist attack, several nations forged a treaty to create a special contingency unit whose main purpose was to establish a squad of ultra-able men and women for implementation of attack and destroy squads. These squads were always on alert if a planned attack to countries of the free world were detected. This special squad unit was called the "No Alternative Attack Unit" (NAAU). The NAAU received several simultaneous alerts from several countries and they met immediately in Geneva, Switzerland.

Their commander, General Edward Magellan, summoned his best teams. This meant alerting special unit teams all over the world. Each team had a different talent and specialty.

All the groups were in the elite category and over-trained in their obligations. General Magellan received his instructions clearly. "Attack and destroy. Do not take prisoners." He activated his drone team then his satellite laser beam specialist group. He then called in the Ninja group and ordered immediate and complete penetration. They proceeded to the order execution point, Havana Cuba. The decision not to use the mini nuclear armament group was made due to Havana's closeness to mainland USA and the possible radiation fallout. Within 12 hours everything was ready, and everyone was in place.

CHAPTER 40

RUSSIA INFORMS

Although Russia and its leaders did not join the Havana group, some of the members of their politburo still had connections with these outcast leaders. The Russian intelligence agency received information as to the intentions and activation of the special group NAAU. They proceeded to inform their ex-ally Castro of the meeting consequences and implications. The Cuban government immediately canceled the meeting and informed the participants of the reasons and the impending consequences of their actions. The Cuban message advised them to be even more careful now that the whole free world had decided to eliminate them.

CHAPTER 41

THE CONTEST ADVANCES: ITS CONSEQUENCES ARE FELT

Time had passed, and our board of directors decided that it was time to close the application process. Some applications continued to trickle in, but we had postponed the deadline so many times that we thought that it was time to stop. The contestant numbers were outstanding. They even included small islands in the Pacific and Atlantic Ocean that were relatively unknown. We even had an application for Point #2 for food production from a group from the Atoll Republic of Maldives a tropical nation in the Indian Ocean. This idea of mine had become a real-world attention getter, or better said, an incredible world-wide attraction. Then early this morning Mike Fredericks called me with the news that was all over the internet, about a secret meeting between rouge country presidents that had been planned in Havana. I called our security people and they had already the news of its cancelation. I wonder what prompted that? Betty called me and asked for a personal face to face meeting, something she had not ever done before. We agreed to meet for lunch. I asked her to invite Mike also.

I received a message from the President of South Korea. He wanted to fly into New York for a private and very important meeting. I told Betty to handle that also. I was overwhelmed with messages and inquires constantly from other country leaders, but it was difficult for me to respond and to meet all of them. I always forwarded all my messages to the office for proper responses. The fact that The Korean president was so emphatic made me accept the meeting. Kathy called back to tell me that I should prepare to meet the Korean for breakfast tomorrow. He asked Betty for total privacy and refused to accept the Korean Consulate as a meeting place. He insinuated that the safest place was your house. He said that he was sure there were no hidden microphones or cameras in your house and what he was to discuss with you was so sensitive that it required total security.

The meeting with my two main collaborators went well. We all noticed the extent of the movement and action that the mere Contest had caused. It had affected the stability of many governments. It had caused many multi-national empires to re-plan and re-think their future goals and objectives. Everyone was putting off plans. Everyone was anxiously waiting for the results. It was not a matter of who was going to be selected as the winner of the contest. Many propositions were going to be presented and most will be valid. In other words, the results would be numerous, and their conclusions were going to revolutionize the world. Nothing was going to be the same. The concept of devising this contest was to stimulate the participants to go beyond their current activities and create a platform for change. We stimulated them with the money giveaway. In reality it was a way of inducing the world to get up and create positive change. The word contest creates creativity and that is why Nelly and I decided to give it that name, a world contest for the good of future generations.

At a glance, the 18 points were going to bring extreme beneficial changes to the world's economies, health, food and water supplies, epidemics, human rights, democracies, and the world problems in general. We might even get rid of the world's extreme climate changes, and we will change the way we travel. Mike was so full of emotion, he told me he had an update on the Nobel Peace Nominations. You my friend, have been nominated by 180 different organizations and societies and universities for the Nobel Peace prize. To add to that, you have been nominated for the Nobel in Economics, Nobel in Chemistry and Physics, and the Nobel in Medicine. The reasons being that your contribution to world peace, world health, and world science has been monumental with your idea of the contest. The incredible results the world expects to receive are simply monumental.

CHAPTER 42

THE SOUTH KOREAN PRESIDENT'S PROPOSAL

At exactly 8 am our house bell rang. My wife went to open the door. She had prepared a quick breakfast and coffee. We did not have help today and this meeting was planned but yesterday. The President of Korea came in. He spoke excellent English, and had his guards wait outside.

He accepted the coffee but said that he had eaten on the airplane. He asked us if the house had been cleared of microphones and we said that the contest company made sure we had no eavesdropping or unsolicited high-technology apparatus installed by anyone. They constantly swept our house to make sure of that. He asked my wife if she could leave, and I told him that I had no secrets from her. She was as much in this as I was. He accepted her presence with uncertainty. I re-assured him that she was 100% security safe. He started talking in a whisper, "I have a man, a specialist, an incredible man that can resolve the problem of the existence of malevolent dictators that oppress so many nations. If we have him liquidate these murderers your contest results will take an incredible leap. Democracies will be immediately restored in all the countries that lack democracies. We can also get rid of the fundamentalist leaders that are the main cause of the existence of terrorism. I assure you that is the only sure way to get rid of these pests, and these known and proven assassins." I felt his immense fury and heartfelt determination to eliminate the worlds scoria. He wanted us to take the law into our own hands. He had lost his emotional control. I understood his request. I was certainly not the correct person to give him my opinion or my approval to such a delicate matter. I attempted to calm him down and explained that many people in this world think that this is the only way of getting rid of these elements. He insisted that he has the right man. "This man will not only do it without harming innocent bystanders but will perform the necessary elimination without ever leaving a trace or evidence. He is a robot, a ninja, he can perform non-human feats. Please tell your contacts to permit me the freedom of action. If the world considers this a crime. I will gladly pay for being the intellectual culprit and planner. I am sure that as soon as the World Court is organized and functioning I will be proclaimed a hero for my idea. Please make it happen." I dismissed him with a greeting of hope. He left with a feeling of accomplishment. He had shared his vision of the only way to resolve the problem, fast, efficiently and irreversibly. He chose to talk to me, because I was instrumental in beginning to change the world.

CHAPTER 43

THE GENIUS ADVISORY BOARD IS FORMED

From the beginning I have thought that we needed a group of genius men and women to come on board and help us keep straight this incredible apparatus that we have been setting up. I know that most of the contestants fall into this category. They were definitely super smart. I started to think of what is the difference between a very smart, bright, individual and a genius? We needed people who had imagination and processing power that could be applied to any situation in our quest for resolving the world's problems. I looked up the characteristics that geniuses share. This is what I found. They have drive. They have a strong desire to work hard and long. They have courage to continue in their project no matter what obstacles come in the way. With devotion and the ability to set goals, they set their minds on their objectives and don't stop until they resolve them. They have knowledge of their inquiry, and they investigate everything that has to do with their subject at hand. They use honesty in their investigation, and they do not even think of changing facts to benefit or speed up their findings. They are optimistic that they will resolve their research satisfactorily. They possess the ability to judge different outcomes and come up with the right decisions. They never lose their enthusiasm to continue without interruption. "Most people have the mistaken idea that geniuses are born, not made", declared clinical psychologist Dr. Alfred Barrious, founder and director of the Self-Programmed Control Center of Los Angeles and author of the book, *Towards Greater Freedom and Happiness*. "But if you look at the lives of the world's greatest geniuses, like Edison, Socrates, da Vinci, Shakespeare and Einstein, you will discover they all had 24 personality characteristics in common. These are traits it is thought can be developed by anyone. It makes no difference how old you are, how much education you have, or what you have accomplished to date. Adopting these personality characteristics enables one to operate on a genius level. Here are the characteristics Dr. Barrious lists, which enable geniuses to come up with and develop new and fruitful ideas: **DRIVE)** Geniuses have a strong desire to work hard and long. They're willing to give all they've got to a project. They develop their drive by focusing on their future success and keep on going until they get to their goal. **COURAGE)** It takes courage to do things others consider impossible. Stop worrying about what people will think if you're different. **DEVOTION TO GOALS)** Geniuses know what they want and go after it. Get control of your life and schedule. Have something specific to accomplish each day. KNOWLEDGE) Geniuses continually accumulate information. Never go to sleep at night without having learned at least one new thing each day. Read. Question people who know more than you. HONESTY) Geniuses are frank, forthright and honest. Take the responsibility for things that go wrong. Be willing to admit, 'I goofed', and learn from your mistakes. OPTIMISM) Geniuses never doubt they will succeed. Deliberately focus your mind on something good coming up. ABILITY TO JUDGE) Try to understand the facts of a situation before you judge. Evaluate things on an opened minded, unprejudiced basis and be willing to change your mind. ENTHUSIASM) Geniuses are so excited about what they are doing, it encourages others to cooperate with them. Really believe that things will turn out well. Don't hold back. WILLINGNESS TO TAKE CHANCES) Overcome your fear of failure. You won't be afraid to take chances once you realize you can learn from your mistakes. DYNAMIC ENERGY) Don't sit and continue waiting for something good to happen. Be determined to make it happen. Make it happen. ENTERPRISE) Geniuses are opportunity seekers. Be willing to take on jobs others won't touch. Never be afraid to try the unknown, it's part of you to do so. PERSUASION) Geniuses know how to motivate people to help them get ahead. You'll find it easy to be persuasive if you believe in what you're doing. OUTGOINGNESS) I've found geniuses able to make friends easily and be easy on their friends. Be a 'booster' not somebody who puts others down. That attitude will win you many valuable friends. ABILITY TO COMMUNICATE) Geniuses are able to effectively get their ideas across to others. Take every opportunity to explain your ideas to others. PATIENCE) Be patient with others most of the time, but always be impatient with yourself. Expect far more of yourself than others. PERCEPTION. Geniuses have their mental radar working full time. Think more of others' needs and wants than you do of your own. PERFECTIONISM) Geniuses cannot tolerate mediocrity, particularly in themselves. Never be easily satisfied with yourself. Always strive to do better. SENSE OF HUMOR) Be willing to laugh at your own expense. Don't take offense when the joke is on you. VERSATILITY) The more things you learn to accomplish, the more confidence you will develop. Don't scare away from new adventures in your field. ADAPTABILITY) Being flexible enables you to adapt to changing circumstances readily. Resist doing things the same old way. Be willing to consider new options. CURIOSITY) An inquisitive, curious mind will help you search out for new information. Don't be afraid to admit you don't know it all. Always ask questions about things you don't understand. INDIVIDUALISM) Do things the way you think they should be done, without fearing disapproval. IDEALISM) Keep your feet on the ground - but have your head in the clouds. Strive to achieve great things, not just for yourself, but for the better of mankind. IMAGINATION) Geniuses know how to think in new combinations and see things from a different perspective better than anyone else. Unclutter your mental environment to develop this type of imagination. Give yourself time each day to daydream, to fantasize, to drift into a dreamy inner life the way you did when you were a child."

A discussion about these traits ensued by the different Board members, and here are the different comments they made.

"Parts of the list are true, but some of them are untrue as well."

"The author seems to have a sort of simplistic outlook."

"I think he is confusing certain qualities as being those of a genius, when in fact, they are actually characteristic of someone who is simply highly intelligent. For example, geniuses are rarely, if ever, so outgoing, optimistic, courageous, or even self-assured. They often suffer from mental and emotional illnesses, are suicidal, isolate themselves, abuse alcohol and drugs, are insecure, too sensitive, high strung from over-worked nervous systems, have personal issues, poor social skills, are misunderstood, have many fears, over-think, are intensely deep-thinkers, have low attention spans and too many interests. Look at Edgar Allen Poe, Isaac Newton, Kurt Cobain, Beethoven, John Nash, Michael Jackson, Van Gogh, Churchill, Hemingway, just to name a few. In fact, name any real genius that you can think of and I can almost guarantee that they struggled with some sort of inner demon or another.

"Highly intelligent people on the other hand are very

outgoing, optimistic, courageous, assertive, self-assured, have strong social skills, good memories, focused and excellent verbal skills. They are quick thinkers, possess natural interpersonal skills, are highly practical, have great judgment and always seem to make the right decisions. These traits aren't Professor Barrious' opinion. He studied "geniuses" and collected data on personality traits they exhibited. He then found patterns. You can always find outliers and exceptions in any data analysis, but this list is the general rule based on his research. The bottom line is these traits are the average found in those people considered genius. Therefore of course you can pick out geniuses that weren't outgoing, or maybe weren't courageous, but if you look at every genius in the history of mankind they are going to exhibit traits on this list. Maybe it is because he has his own personal agenda to uphold and is thus selective (whether consciously or unconsciously) about the "data" in order to support his own *a priori* definition of what a genius is (a possibility which doesn't seem too far-fetched if one considers the fact that he believes that genius can be achieved and need not be innate.)" If the author is right, then why is it that I can't think of a single genius that fits his description? Maybe I'm just not well-read enough, but almost every known genius that I can think of seems to match Joe Blow's description much better (and I think most people would seem to agree)." "Some other examples of the tortured genius include Nikola Tesla (obsessive-compulsive disorder), Stephen King (former alcoholic and drug addict), Stephen Hawking (has a paralyzing motor neuron disease),

George Trepal (anti-social, poisoned his neighbors), Michael Crichton (social outcast and loner), Abraham Lincoln (major depression, considered suicide), Edvard Munch (panic attacks, anxiety disorder), Michelangelo (autistic), and the list goes on and on. And consider Dr. Philippe Rushton's research on the correlation between race and intelligence in which he argues that East Asians are, on average, the most consistently intelligent of all the races, due to their large brains, low body mass, low production of testosterone (which is detrimental to brain development). Is it a mere coincidence that East Asians tend to be shyer, introverted, quiet, reserved, introspective, cautious, self-conscious, precise, and intuitive, laidback and confident? There are way too many exceptions to the rule of what constitutes a genius, what a genius must have, how they must spell, think, or act. I feel myself as highly gifted from my earliest thoughts on this planet. I was exposed to people full of hate or ignorance almost daily. Am I a genius? Am I highly gifted? Am I Eccentric? I can tell you that the reason some geniuses go down a dark path is the environment and idiots they have to deal with. So supportive factors are not there for these highly intelligent people. There are far too many people who are idiots that are in leadership positions that are trying to squash these brilliant people. I have experienced it firsthand my entire life. They are beginning to recognize this more and more, and cultural change will happen, eventually. Thanks to the Internet the issues, discussions, the cyberbullying attacks, etc., etc. get out much quicker... Change is coming... It is shameful that the people of the world have had to suffer at the hands of evil, greedy, controlling people, leaders etc.... Only because these people are jealous, have degrees maybe higher than the genius and feel better than them. Elvis probably wasn't a genius ...but if he was he would fit your mold ...so would Greta Garbo...but William F. Buckley Jr. was a genius in my opinion and does not fit the neurotic tortured model. Then, we have a definite link with geniuses and winners of the Noble Prize. From the historical information we notice that almost all Nobel winners were achievers beyond the normal limits of human knowledge. Take for example Albert Einstein, a very slow learner who was even kicked out of one of his schools as a sort of underachiever with a rebellious nature. He was considered a nerd of sorts, always distracted. This trait of slow verbal development might have allowed him to observe everyday phenomena that others did not observe. No one seemed to wonder about the problems of space and time. Einstein revolutionized the world when he came up with the concepts of the theory of relativity and quantum theory. These two theories are still valid today. We can mention many with genius-like attributes who definitely made a difference in the world. Marie Curie, the first woman to win a Nobel Prize, really won two. One for 'The invisible atomic particles that challenged the laws of thermodynamics, and the second for her work on radium and polonium which opened the world to the use and cure with radiation. Well, your ideas are interesting, let me tell you what I think. I just thought of Leonardo da Vinci, he had the ability to visualize theoretical concepts. in my opinion, he was history's greatest creative genius. He did not have the theoretical brainpower of Einstein or Newton, but he created incredible works of art, incredible tri dimensional maps, and memorable drawings. da Vinci also had a problem with authority, something that many geniuses have. Could this criterion apply to most? da Vinci was self-taught and considered himself as a disciple of experience. His biggest asset seemed to be his indefatigable curiosity. His curiosity extended to all areas, he wanted to know why we yawn, what makes an aortic valve close, he studied about the placenta of a calf, the jaw of an alligator, the light of the moon and even the description of the tongue of a woodpecker. He wanted to know everything about everything. These observations and questioning of everything led him to create some of the most brilliant art in history." After the discussion ended we knew exactly what our contestants' principal characteristics were. They were all either geniuses or very intelligent, motivated people who wanted to change the world by doing their part.

CHAPTER 44

THINGS BECOME DIFFICULT

While all the planning and all the different contestants were on top of their projects, I received an email from a schoolmate that I had corresponded with, on and off during most of my life. He said he needed to speak to me about something that he had read on the Internet. He said that it was important, and that it had to do with my project, the World Contest. The fact that he had used my private e-mail, and the fact that I knew the man as a serious individual, prompted me to answer him and set up a telephone meeting the next day. I couldn't sleep that night thinking of what he was going to tell me that was so important. Nelly, my wife, kept telling me to forget the telephone meeting with Philip Poster. She said that he was probably trying to get close to us now that we were in the news every day. However, my curiosity drove me to call him before our scheduled time. I hadn't spoken to Philip in at least 20 years. He was an online marketeer and was into importing gadgets from China and other

Asian countries. His voice was immediately recognizable. He congratulated me and then went immediately into the subject of his email. "Look buddy, I have some un-quieting news. While searching the dark web, I discovered that you and your organization are mentioned all over the place, and all indicates that you might have a major problem in the near future." "What is the dark web?" "We have the regular web that you and everyone knows about and then, we have the dark web. This is a sub-set of the deep web that contains all sorts of websites, both legal and illegal. The types of sites include black markets that sell things like drugs, counterfeit goods, and weapons. You also have hacking sites, porn sites, bitcoin tumbling, and even sites for hitmen. The variety of sites on the dark web is quite astonishing. I sometimes go there to see the Bitcoin transactions. When I went in yesterday morning, after a few minutes I saw reference to your organization, and your name is all over the place. Someone has put a price on your head. Someone wants to eliminate you and everyone in the World Contest. They have offered 50 million Euros or its equivalency in bitcoins for the job. I immediately asked him to send me the website or web location. He hesitated and said, "I do not want to get involved, have your people search that area and they will find it immediately, I am sure." I thanked him for the information and told him that I would get back to him. He asked me to keep him out of this and, please don't call me. "Please do not tell anyone that I was the one that warned you, I do not want to get killed." Nelly was listening attentively to our conversation and she became altered, by what she heard. For the first time since we started this incredible adventure she started crying and told me, "I suspected that something bad was going to result from this. Let's get out of this. This is not for us. We are too old for this type of life, for the dangers that I now hear we are involved in. This world is full of crazies." I got on the phone and called Mike and told him everything that I could without mentioning Philip. He told me not to worry and to continue with my life as usual. He said that we already had surveillance attached to our every movement and he would warn the detail in charge of us. He would have this investigated thoroughly and advised me not to believe everything that was on the dark web.

CHAPTER 45

INTERPOL IS CALLED

Mike called his security people and they in turn decided to involve every United States security governmental and international police institutions in this very scary affair. At the end everyone recommended that the Interpol should handle it because too many organizations doing the same thing was going to be counterproductive.

Interpol is The International Criminal Police Organization. What it really does is facilitate international police cooperation. Its headquarters are in Lyon, France, and it's been around since 1923. 192 countries belong to the Interpol. It is the best mechanism the world law enforcement has, to coordinate and arrest wanted criminals that happen to be in a foreign country. Only very few countries do not belong to the Interpol, the most striking example is North Korea.

Once the FBI sent the researched information to the Interpol, all the member countries were alerted and were asked to intervene in this affair. A French inspector was assigned to the case. Inspector Raphael Arcand was a very experienced forty-year-old police officer. He was once in the French Foreign Legion and was highly decorated for incredible secret accomplishments in favor of world peace. He had contacts in the underworld in Algeria, Libya, Tunisia and other African countries. His principal aide was a web hacking specialist, who graduated from Georgia Tech computer science dept. He put into motion an intense search for the culprits of the incitement and searched for the paid assassins on the dark web with its millions of sites. Within hours he received a note from a Tunisian operative with confidential information for his ears only. He implicated a North Korean group digging for uranium in an abandoned mine near the Algerian border. There was another group in a mine near the Libyan border. Here is where the width of the land tapers into the south-west and then into the Saharan desert. Both groups were in continuous communication and used the dark web as the tool. This was probably the best way to remain undetected, and that's why they used that web site. Inspector Arcand was given the use of a large airplane that carried, a top of the line 4-wheel drive Jeep, with all kind of high tech communication gadgets installed. He also enlisted one of the women in his department whose duty was to attract and distract the potential enemy and obtain the necessary information. She was also a disposal specialist. She had the license to kill if necessary. She had already been decorated over 50 times for her valor and successful completion of specific assignments in many parts of the world. Her main attribute was her beauty. She was usually disguised, since she was also a disguise specialist. Her name Agustina de Pepina, originally from Pescara Italy.

Detective Arcand and his group of volunteers arrived at Carthage International Airport very early in the morning. The Tunisian police liaison was awaiting them. They were exempted from going through the normal immigration and customs procedures. They dispensed with small talk and unpacked their special Jeep, loaded the special communications equipment and followed the Tunisian vehicle to their destination, the closest town to the Libyan border. Inspector Arcand had a feeling that was where the head of the Korean group was hiding. Agustina jumped into a Tunisian Police helicopter and headed to the same location to prepare for the eventual meeting with the North Koreans.

Agustina arrived in Ben Guerdane, the closest town. She checked into a hostel, the only one in the area. A few minutes later she emerged dressed as a Tunisian Army officer. Since she spoke French as well as Arabic, she was able to communicate and hire a Bedouin Taxi driver for the short trip to the Korean mining post. Everyone in the area was armed. Few if any women were visible. There was fear in the atmosphere. She was told of a rumor that the Isis group had been seen in the border with Libya. The Bedouin told her that over 30 men had been accused of being unholy and beheaded at the Libyan border. Her Bedouin guide spoke to Agustina with respect because she was armed and evidently ready for any confrontation.

Agustina reached a point guarded by a Tunisian army unit. Once they saw Agustina they immediately saluted her due to the evident high rank. She was a superior officer and she responded without talking. When she reached the North Korean guards, she spoke to them in French, and the head of the post a tall, thin, ugly fellow came forward and asked who she was and what she wanted. She told him that "I was sent to inspect and report on your activities in this mine." The Koreans immediately started shouting and she took out her handgun and grabbed him by his neck and silently said, "You are in my country and if you don't like it you are out of here immediately." He had been overpowered with such force, he felt that he'd better obey. He could not afford to have a problem with the Tunisian Army. Agustina motioned to the Bedouin taxi driver to accompany her and one of the Tunisian soldiers to protect her back. She told the tall skinny one, "I want to meet with all of your team at 5 pm. sharp. I want to see the minerals that you have ready to ship out of the country and I want to see how you pack them. I want to see the paperwork of everything shipped in the last 6 months." The Korean protested again, and she again warned him that, you obey, or you leave today. She went into what seemed an antenna and communication room and left the guard outside with instructions. No one enters this room without my permission. She then went into the food canteen and served herself some of the fruit that was available. The Korean was extremely upset and didn't know what to do. He had to report this to his boss. This Tunisian lady officer seemed to be receiving instructions from the head of the Tunisian Military. He excused himself and went into the improvised latrine. Once there he quietly texted his boss of the presence of trouble with a military colonel. The boss texted back, I'll be there in four hours. At exactly 7 pm. Inspector Arcand arrived in the small town. He and his men all dressed in Tunisian army clothing arranged their equipment in one of the small hostel rooms and communicated with Agustina. Agustina was checking the Korean group and their recently mined uranium ore. There were over 40 men, no women. The inspector had her return to the hostel immediately and she told the Tunisian soldier to leave the communication room and come with her. In less than two minutes the Inspector's specialist noted that the dark web started to transmit information as to the presence of intruders in their camp. They used the same frequency and the same web connection units that had been identified as the ones seeking the assassination of the World Contest organizers. We did not have sufficient information, we didn't know who the culprit was. We think that the transmission came from this Korean communication room.

Agustina explained everything to the Inspector. She had done an excellent job. She was now going to make an appearance as a seductress. She dressed appropriately as an Arab woman, but underneath her gown, she was ready to conquer even the most hardened and difficult man. She had a body that no normal man could resist.

The Korean boss arrived at around 7 pm and after being

put up to date, he asked for the army officer that had visited the camp. He was informed that she had returned to the city. The inspector was in the makeshift coffee house sitting with his men, when the Korean approached them. He also spoke perfect French and the Inspector greeted him. The Korean said that he was going to lodge an official complaint because of the intromission of the Tunisian officer in their totally permissible and paid for contract regulations. The Inspector excused himself and denied having knowledge of the occurrence. That's when Agustina came out accompanied by her Bedouin taxi driver, now dressed as an Arabian warlord. She had really done a job on him and herself and even the Inspector and his men did not realize it was Agustina. The warlord had been instructed to make a scene after consumption of hashish and he was told to openly mistreat Agustina, in public. The idea was that the Korean should react and help Agustina. The Korean ignored all the swearing and improper offenses toward the Arab woman. So, Agustina ran toward the Korean for help. She unhooked her gown sufficiently for him to observe a beautiful pair of breasts and she had no bra. That changed everything, the Korean got up, pushed the Bedouin to the floor and saved Agustina with a protective hug. She clasped him and embraced him. She Let him feel the warmth of her lips on his neck. That did it. He spoke to her in Arabic and she accepted. He went out of the coffee shop with her. They went straight to the Korean campgrounds. Agustina clung to him and even though he did not want to be seen by his men, she didn't let him separate from her. They went straight to a room full of beds. He had everyone vacate the room. She jumped on top of him and, he was in heaven. He couldn't believe his luck. He had not been with a woman in over 3 years, since he left his village back in North Korea. This woman was beautiful, and she was very intense in her objective. When thinking of her motive he assumed it was to thank him for saving her from that brute. When he undressed, she started to speak to him in French. She told him that she would have sex with him but that she wanted to get married first. She said she was a virgin and her family would kill her if she had sex before she married. He told her he couldn't marry her. I am married, and in my country, I can go to jail for getting married twice. She attempted to convince him that he didn't have to return. She told him that she was sure that they had sufficient Uranium here, they could get very rich with it and move to France. She let him see more of her naked parts, these drove him crazy. She insisted, he melted to her wishes. She got dressed, he got dressed in frustration and told her that he would do whatever she wished. He was in love with her. She told him that she had to go now but would meet him later. She left, and on the way out she saw the tall, thin Korean. He asked her who she was and what she was doing here? She answered in Arabic, "I am lost, can you accompany me to the town" She showed him enough of her body for him to become interested. He followed her, and she made the trip seem worth his while, with continuous insinuations and sensual movements of her body. He was hypnotized. He followed her up to her room and she started to interrogate him, she showed him a little bit more of what he wanted to see until she knew everything she needed to know. She made a signal, and Inspector Arcand came into the room. The mission was almost accomplished. The Korean who was mesmerized came back to reality, and was asked to get dressed, and leave the room. They all met in the coffee house. The Inspector had brought several bottles of Chateauneuf du pape with him to celebrate victory. Inspector Arcand wrote a very detailed report back to Interpol headquarters.

"The Dark Web is being used by North Korea for their private communications in Tunisia. Since they are in an Arab guerrilla area they have permitted the use of their communication setup by Libyan-led fundamentalist groups that have been trying to upset everything in the free world with threats and attempts to hire assassins through this obscure dark web interface. We will destroy all the communication and installations here, if you so command, before leaving tomorrow." The response was immediate. "Do whatever is necessary to eliminate evidence of your presence."

At exactly 4 am the Korean compound was totally destroyed. There would be no more communication or uranium mining for a long time. We felt sorry for the Koreans, and we let them live. They would probably be killed if they returned to North Korea. The Tunisians that helped them were promised a monetary bonus. The Bedouin Taxi driver got a double bonus and they let him keep their special Jeep.

Inspector Arcand was decorated again, this time by the World Contest organizers. The decoration should have gone to the real hero, Agustina.

CHAPTER 46

MIKE DISAPPEARS

Mike had received the good news that the dark web perpetrators had been neutralized and he immediately communicated this information to Nelly and me. Several hours later we got a call from his office, asking us if we had heard from him. Mike had missed his lunch appointment with the President of Jamaica, and his 2 o'clock dental appointment also. There were two people with appointments waiting for him. I told them to call Betty. They already had. I then became worried. I sent someone to his house, and he wasn't there. His security guard said that he had given him the day off, something very strange. After two hours of trying to locate him were not successful, Betty came over very sad. We had called all the hospitals and no luck. We notified the police and they started posting his pictures to all precincts. That night we hardly slept. The next morning Betty received a message to her phone. Someone was asking for ransom money. Mike had been kidnapped. We alerted all our security systems again. When inspector Raphael Arcand was informed of the occurrence, he asked his superior officer for permission to investigate this case. He had read everything that the world Contest was doing, and he really wanted to participate in their incredible quest to help change the world. His superior officer accepted the inspectors wishes and also gave the go ahead to Agustina, the inspector's right-hand woman. They were well known in police circles and had special diplomatic passports that did not require a visa to any of the Interpol member nations. That evening they both left on a direct flight to Kennedy airport. They were greeted by the US Interpol representative that had arranged all their necessities.

Within the hour, both were in Betty's office and discussed a plan that both inspector and Agustina had made during the 10-hour flight. Betty had received the phone call informing them that they needed to meet with her to explain the demands they had, in order to permit Mike to continue with life. She would be receiving a call in the next hour. They decided to stay in the office together with a large team of high technology gurus in attempt to intercept the whereabouts of the impending callers. After hours of impatient wait, the call came in to Betty's private land line. Agustina waived Betty away from the phone, and after they let it ring 5 times, Agustina picked up the phone. An apparently disguised and recorded voice instructed Betty to come alone to a very specific area close to the dock area of the Jacob Javits convention center. The instructions were clear, if you are followed,

Mike is dead. If we suspect any monkey business on your part Mike is dead. We are in this no matter what happens to us. Our objective is either we succeed, or we die. We warn you be there at 11 pm sharp, if you are late Mike is dead. The recording ended, she then hung up. The specialist could not identify the origins of the call. One thing they knew for sure, is that it originated in Central Asia. The recording was dubbed to simulate an Australian accent.

Agustina immediately accompanied Betty to her apartment and began the process of disguising herself in the mirror image of Betty. Agustina was also a very good voice imitator and by now could perfectly imitate Betty's voice. She went into the bathroom, and minutes later came out looking like Betty's identical twin. Betty's astonishment could easily be noticed on her facial expression of disbelief. First, she went out into the dining room where some of Betty's aides were nervously talking and Agustina started conversing with them. To the amazement of everyone the real Betty came out a few minutes later and introduced herself as Betty also. They could not tell who the real one was. Their voice, their manners, their way of walking, everything, was identical. They all received instructions to carefully disperse, in case the apartment building was being watched. Although this building was gigantic and had multiple entrances, they did not want to take a chance. Betty came out dressed like a male assistant and left with her aides. Agustina, left the building alone, went to a nearby late-night coffee shop, sat down ordered a coffee and waited for the right hour. At 10;40pm she called for an Uber. The car pulled up 3 minutes later. She got in to the small car, the driver seemed to be acting strangely. He did not even greet her, which seemed very strange, because they always ask you for your name. Then, instead of heading toward the Javits Center, he hurriedly made a left turn into an ally and someone opened the right-side door and grabbed Agustina and placed a head covering on her. No one spoke, Agustina senses started to operate. She knew what she had to do. She remained calm but for appearance sake acted as totally terrorized with fear. They changed cars three times, and then she was taken to a boat, she could not tell if it was a small or large one, they then placed ear plugs under her headmask, so she could not hear their conversations or the nearby noises. The boat stopped, she was taken up a staircase of a larger boat, she assumed that because of a familiar smell that she recognized as that of a cargo ship environment. After an hour or so they must have used chloroform on her, because she woke up and saw a slight ray of sun pierce thru the improvised blindfold. She knew by now that they would not harm her and let her live. They had taken every precaution not to be discovered. They had made sure she had no transmitters. They had done a complete and total search, a female was involved, because someone invaded her private parts and they were certainly not masculine, during that search.

Inspector Arcand had taken a necessary rest and, his Interpol assistant, woke him up when they heard that Agustina was going to be questioned. The inspector had informed his men that Agustina had surgically been implanted with a very small and sophisticated plastic implant inside one of her breasts, many years ago. They had recently changed the re-chargeable battery and it had the unique characteristic that it remained totally inactive until the inspector activated it from a specific distance with a world satellite password procedure.

Agustina was feeling the now familiar mini-electrical current of waves that she felt when she was being followed or scanned by the inspector or his aides. The breast implant had been activated. She felt safe and knew that if necessary she could ask for assistance wherever she was taken by pressing her right breast in a certain way. She was permitted to take off the mask and earplugs to find herself in a barren room, except for a single bunk bed, a toilet can, and a tray with some hot oatmeal and a glass of water. She decided to consume both, she needed the energy, and she knew that it was unlikely that they would put her to sleep again. An hour later, the lights were shut off again, and someone came in and put a mask on her face, but not the earplugs. Then she was taken to an air-conditioned room. She felt the presence of two or three persons, because of their intermittent breathing. They were all nervous, she could tell, as she was trained in these matters. Agustina decided to act out a scene and started to cry and complain of the torture they were putting her through. She wanted to psychologically provoke them, attempt to create pity, or anger in the guys she felt breathing. Then she heard the door open and a woman came in. She was definitely in command, the others in the room became even more nervous with her sudden entrance. She spoke with a Russian accent; her English was good. She must have been tall. Agustina analyzed her every move and word. "I hope my men did not treat you badly, I know you very well. We met many years ago when you were the President of the World Bank, but I'm sure you don't remember, you were famous then. You are famous now because of the stupid Contest that you are involved in. You and your lover Mike have to make it disappear immediately. You are affecting many world leaders with your stupid solution finder. You are destabilizing many organized countries in the world. My stupid men captured the wrong man. They needed to bring Dr. Cantor and his wife, and they brought you and Mike. However, the fact that we have captured both of you will probably make it easier for us to deal directly with Cantor, the promoter of all these absurd contests. Call me Angelina, I will make your life miserable, and I will have your lover undergo horrible moments, so you will wish that both of you were dead. I will let you go, and I give you one week to find and implement a way for the Contest to auto-destroy itself. If you do not achieve this, or if you try to trick me in any way, I will send you pictures of Mike's hands, then his eyes, as I have them removed one by one." As Agustina listened to her terror threats, she discovered that her mask had a little tear on one angle and saw the so-called Angelina's left profile and then her front profile and she recognized her. Angelina was a spokeswoman for a terrorist organization in Yemen. She knew that she had to remain quiet and obedient. Agustina then spoke and told her that she needed proof that Mike was alive and well. She became nasty and told her that, "You are in no position to ask for anything. If you do not obey me your boyfriend will cease to exist." Agustina said, "Angelina please understand, I need to see Mike, if I don't, kill me, I will not destroy the Contest. I'm in love with him. You are a woman you must know what that means." "Shut up, don't tell me what to do," is what she heard from her mouth. Then she received a phone call and Agustina could hear her breath become very loud, and she approached her and started to feel her legs and her behind, and then her breasts. She was unto something. She had gotten a call warning her that she might have a tell-a tale-chip somewhere in my body. "Or was she a lesbian enjoying my body?"

CHAPTER 47

THE INSPECTOR MAKES A DECISION

Inspector Raphael Arcand listened attentively as Agustina was being threatened or instructed, by the so-called Angelina. Then all of a sudden, he detected motion in the trigger chip that had been implanted on Agustina's breast. He knew it was not being manipulated by Agustina, who he was sure had her hands in handcuffs or another appliance that did not permit her hands freedom of movement. Was Agustina being checked for a transmission chip? He could not tell. He could hear that her respiratory wave lengths had increased twofold. Was she in danger? He could not act hastily, because the whole operation could be jeopardized. He pinpointed the location on a map. They were on a ship, 12 miles off the coast of New York in International waters. He called his contact in the United States Navy and advised him that he might need a helicopter backup in the New York area for an immediate agent recuperation deployment. He told them that the action will be performed by authorized

Interpol agents. The favor was for the Navy to supply a helicopter and an experienced crew for transport only. The Navy contact said he would immediately ask for permission and have a backup in case something went astray.

Meanwhile, Agustina continued to be hand handled and complained to her female captor. Something was definitely wrong. This Angelina was having a weird type of satisfaction. She did not know whether to accept being touched, or somehow free herself from the handcuffs and finish this charade. She knew that whatever she did could cause this animal of a woman to cause harm to Mike." "Angelina, what are you doing?" "Are you searching me or feeling me up? Let me know, if you are feeling me up untie me, let me enjoy myself also." With this, an incredible thing happened. Angelina told the guards to take off her handcuffs. They were amazed at the order and did so. She then instructed them to leave her alone with the captive. They left the room. Agustina got up relaxed her hurting wrists and took off part of her blouse to expose her breasts to the more than excited Angelina. This action drove Angelina into a strange situation, into a frenzy. She started to undress. Agustina seized the moment to grasp her throat and placed her into a lock position where she could hardly breath. She quietly asked her, "Where is Mike or you are dead" It didn't take long for her to admit that Mike was on the ship. Agustina pressed her breast and said out loud, "Get here as soon as possible." The message was very clear, the Navy Helicopter was three minutes from the cargo ship, and 20 men were ready to board. It took less than five minutes to find Mike. Agustina had tied up Angelina with her own clothing and proceeded to move fast to aid the men she knew were about to arrive. To everyone's surprise there were only six men on the ship. Inspector Arcand directed the ship into United States waters. The boat had been leased by a Libyan businessman from Nigeria. Now that the ship was not in International water, the US Coast Guard arrived and took over the police investigation. Mike and Agustina were airlifted back to Manhattan.

Dr Cantor, Betty and the complete group were there to greet them very relieved with the happy ending. That night Inspector Arcand, Agustina, and the whole Interpol group were given a special tribute for their phenomenal performance.

CHAPTER 48

THE CONTEST IS UNSTOPPABLE

A lthough we have definitely met political and terrorist led obstacles, to name just a few, nothing has been able to stop us. Time advances at a fast pace. The whole machinery of the contest is totally functional. Many of the different contestant groups have presented their preliminary findings to our newly appointed Board of Geniuses for their review. This step was a requirement for approval of additional funds. The Climate control group was really up to something. This was not the only group. In the last week we were informed that almost all the groups had requested appointments for review of their project. In other words, they had advanced and could prove their performance and wanted additional funds to continue or to complete whatever they were attempting. One extraordinary happening was the very close cooperation from the Chinese and the Russian board members. They totally integrated into our way of doing things the democratic way. It was a great beginning for the change we wanted to project for the future. All the countries in the world were getting involved to improve their own and the world's economy, health, food availability, weather control, elections, democracy, equality in education, and so much more.

CHAPTER 49

DR CANTOR'S THOUGHTS

I got up early this Sunday and I looked at my calendar. Today was exactly the third month since I discovered that I had won the biggest prize ever paid by a lottery in the world. That was the day that our life changed forever. The life of our children and our close relatives and close friends changed forever also. We couldn't stroll down the mall and meet for coffee informally near Nordstrom's cafeteria. We couldn't go to a movie or shows or to a restaurant unaccompanied. We needed guards and security details wherever we went. Is this the life that we wanted to lead? I'm not sure. Are these the consequences of becoming famous? What do we have to do to recapture our tranquility? Would we have to travel to a small and unpopulated island for privacy?

Betty came in and embraced me. She told me that she loved me. She told me that I needed to go to a Dr Young. "He might help you get over these hallucinations that you have been having lately. Maybe if we really win the lottery we can start to travel."

Made in the USA
Columbia, SC
10 May 2022